SpringerBriefs in Physics

More information about this series at http://www.springer.com/series/8902

Ignazio Licata · Leonardo Chiatti
Elmo Benedetto

De Sitter Projective Relativity

 Springer

Ignazio Licata
ISEM (Institute for Scientific Methodology)
Palermo
Italy

Elmo Benedetto
Department of Engineering
Università degli Studi del Sannio
Benevento
Italy

Leonardo Chiatti
AUSL Medical Physics Laboratory
Viterbo
Italy

ISSN 2191-5423
SpringerBriefs in Physics
ISBN 978-3-319-52270-8
DOI 10.1007/978-3-319-52271-5

ISSN 2191-5431   (electronic)

ISBN 978-3-319-52271-5   (eBook)

Library of Congress Control Number: 2017933439

Printed on acid-free paper

This Springer imprint is published by Springer Nature
The registered company is Springer International Publishing AG
The registered company address is: Gewerbestrasse 11, 6330 Cham, Switzerland

The authors dedicate this modest work to the memory of distinguished professors Luigi Fantappié and Giuseppe Arcidiacono

Luigi Fantappié; Viterbo (Italy) 1901–1956

Giuseppe Arcidiacono; Acireale (Italy) 1927–Rome 1998

Prof. Giuseppe Arcidiacono and Ignazio Licata (left) converse during a break in a scientific meeting (1991)

Foreword

Half a century ago, the proper name "de Sitter" was known only, or almost so, to a small bunch of theoreticians working in cosmology—which was far from being such a fashionable and active subject as it is now. De Sitter had in fact discovered a few decades before a particular solution of the Einstein's equations in general relativity, describing an expanding universe, unfortunately empty of matter and thus appearing to lack a deep physical interest. However in the 60's, as the group-theoretical foundations of Einsteinian "special" relativity were becoming common knowledge in mathematical physics, the attention was drawn to the symmetry properties of the de Sitter universe, expressed through a specific group structure generalizing the Poincaré group which underlies the fabric of the Minkowski space-time.

It is revealing that, while, up to that period, the name de Sitter appeared essentially in the expression "de Sitter universe", new syntagms such as "de Sitter space" and "de Sitter group" quickly spread thereafter. These mathematical objects were then studied in their own right and shown to possess some very elegant properties, which made them an interesting playing ground for quite a number of works dealing for instance with the problems of quantization in a curved space-time as well as, of course, with cosmological considerations, especially when the cosmological constant came back into favour.

Nevertheless, as pointed out by Dyson in 1972 and recalled in the introduction of this book below, a serious examination of what Dyson called "de Sitter relativity", expressing the chronogeometry associated with the intrinsic structure of the de Sitter space-time, was still lacking. Or so it seemed. For, indeed, as revealed by the authors of the present book, an abundant and thorough work on the subject had been done in the 50's by the Italian mathematician L. Fantappiè followed by his colleague G. Arcidiacono. Unfortunately, their papers were mainly written in Italian and published in mathematical journals, which accounts for the fact that they were largely ignored, another reason probably being that they did not refer explicitly, at least in their titles, to de Sitter.

Beyond the case in point, there is here a rather general lesson to be drawn. Out the immediate reach of main line research and escaping the linguistic domination of basic English in hard sciences, there certainly lie quite a number of interesting and promising studies in many domains, buried in high quality but low readership journals in various languages, sometimes going back to an almost forgotten past. The present electronic information systems are quite unadapted to tracing back such works of value. One can only wish that brave young princes of science will try to discover and awaken these sleeping beauties.

Coming back to the specific theme of this book, one must be grateful to the authors for this timely review which has the double merit of giving due regard to the work of unjustly forgotten precursors while at the same time offering a thorough perspective on modern developments.

At this very day, the interesting figure of Willem de Sitter and the wide scope of his scientific work remains largely ignored. It is thus worthwhile to devote a few lines to the man.

Willem de Sitter (1872–1934) was a Dutch astronomer with a brilliant career. While he had a mostly mathematical training, he was at the beginning interested in celestial mechanics. He did some observational work at the Cape Observatory in South Africa, where he went in 1897–1899, in order to, according to his own words, "complete my astronomical education—or rather begin it, for up to that time I had never made a speciality of astronomy and intended to become a mathematician". His doctorate and most of his early publications in the first decade of the twentieth century were dedicated to a thorough analysis of the motions of the planet Jupiter and its satellites. In 1908, he took up the chair of astronomy at the University of Leiden, and in 1919, was appointed Director of the Leiden Observatory in addition to his professorship. At that time, he had started to work on the then very recent theory of general relativity, which he was certainly one of the first to understand in depth. He wrote an early non-technical presentation of general relativity still worth reading ("Space, Time, and Gravitation", *The Observatory*, n° 505, 1916, 412–419, to be found online at http://articles.adsabs.harvard.edu/), and went on to publish a number of significant papers on its astronomical consequences. De Sitter, unlike Einstein, stressed and maintained that general relativity actually implied the expansion of the universe. In 1932 Einstein and de Sitter published a joint paper in which they proposed a simple solution of the field equations of general relativity for an expanding universe. They argued in this paper that there might be large amounts of invisible matter. This was the first hint of the now well-known but still mysterious 'dark matter'. Although de Sitter is best known for this work on relativity, he made many other contributions of great significance. He kept his interest in Jupiter's satellites throughout his life, using data on eclipses of the satellites dating back to 1668 in order to produce definitive results on the orbital elements and masses of the four Galilean satellites, on which he was still working when he died in 1934. Another study which de Sitter undertook was to refine the data for the fundamental constants of astronomy, associated with precession, nutation, solar parallax, lunar parallax and the mass of the moon. At the time of his death, de Sitter had almost completed a new updating of these constants.

De Sitter was much appreciated and admired, as shown by the following extract of one of his obituaries :

In [de Sitter's] work we see the creative mathematician at his best. He is not a cold, dispassionate juggler of Greek letters, a balancer of equations, but rather an artist in whom wild flights of the imagination are restrained by the formalism of a symbolic language and the evidence of observation. Only the musician can fully grasp what it must have meant to de Sitter to see the cosmos shaping itself in new ways in his formulas. Like musical notes, strange symbols, standing for forces and masses that were divined rather than known, arranged themselves into a coherent message. And when the message came to be read a totally new universe was revealed. Here we have something of the direct personal experience of the outer world, of the significance of nature's wonders, that comes only to a Beethoven or a Milton. The expanding universe of de Sitter must be regarded as something more than an inexorable conclusion drawn from the strictest kind of logic with which the human mind is familiar. It is poetry of a new sort - the scientist's way of writing an epic.

October 2016   Jean-Marc Lévy-Leblond
                                              University of Nice Sophia Antipolis

Preface

The idea of this little book was born many years ago when, just graduated, Leonardo Chiatti and I frequented Giuseppe Arcidiacono.

Furthermore, Leonardo lives and works in Viterbo, Fantappiè birthplace. I remember our long bus tour through Rome while talking about Einstein and de Sitter. The first time I met Arcidiacono is set in my memory indelibly. I was doing military service in L'Aquila, and after a long exchange of mails I made an appointment with the Professor at his home in Rome. As soon as he opened the door—avoiding any formality!—asked me what I thought about Lorentz Symmetry! And we went on in this way for about 4 hrs, then he opened the old box of Sicilian pastries where he usually put the mails by Freeman Dyson, Jean Marc Levy Leblond, Feza Gursey, Erasmo Recami and some other names the reader will meet in the pages of this book. The Euclidean line of attack to Quantum Cosmology by Hartle and Hawking stimulated Leonardo and me to recover the Group Approach. We realized that a long work of revision had to be done before the "quantum jump" ahead. In fact, most of the papers were in Italian and followed a classical approach. We, finally, completed this work in 2009, too late to be seen by Arcidiacono.

Later, a young and brilliant mathematician, Elmo Benedetto, joined us. We all decided to give the reader an agile and unitary vision of de Sitter projective approach, an extremely fecund one in particle physics and cosmology.

This is a little book, but with big debts: first of all to two great masters: G. Arcidiacono and L. Fantappié. And then to a formidable group of colleagues and friends: Eliano Pessa, Erasmo Recami, Giuseppe Vitiello, Jean Marc Levy Leblond, Basil Hiley, Jose Geraldo Pereira, Reuben Aldrovandi, Ugo Moschella and PCW Davies.

The book is dedicated to the masters, but more to new generations of physicists and mathematicians.

Palermo, Italy   Ignazio Licata
Viterbo, Italy   Leonardo Chiatti
Benevento, Italy   Elmo Benedetto
December 2016

Contents

Chapter 1
De Sitter Relativity: A Sixty-Year-Long Story

In this chapter we sketch out a historical overview of de Sitter Relativity, focusing on the projective version developed by the Italian mathematicians L. Fantappiè and G. Arcidiacono in the early 1950s. In addition, we analyze tendencies and problems connected to de Sitter Relativity in contemporary physics.

1.1 Introduction

This book proposes to provide a short and concise modern introduction to what is known as the "de Sitter theory of relativity", setting out its underlying concepts and briefly touching on the implications having a greater impact on current physics research, especially in the field of cosmology. All that is asked of the reader is a basic acquaintance with special and general relativity and the rudiments of cosmology.

As will be explained further on in this chapter, de Sitter relativity has been rediscovered independently several times by different authors and years apart from each other; consequently, the literature offers different presentations and also different physical interpretations of its formalism. We shall follow the original formulation (the so-called "projective" one) developed by L. Fantappié (1901–1956) and later by G. Arcidiacono (1927–1998).

We are actually dealing with two distinct theories: the projective special relativity theory (PSR) and the projective general relativity theory (PGR). The former is a generalization of the ordinary theory of special relativity (SR); the latter is the corresponding generalization of the ordinary theory of general relativity (GR). The relation between PGR and PSR is the same as that between GR and SR. We shall then discuss about "projective relativity" or "de Sitter relativity" (dSR) and refer, as circumstances require, to PSR or to PGR. Naturally, "projective relativity" in this sense is not to be confused with the theory that goes by the same name, but is

© The Author(s) 2017
I. Licata et al., *De Sitter Projective Relativity*, SpringerBriefs in Physics,
DOI 10.1007/978-3-319-52271-5_1

entirely different, formulated by Veblen. The difference between the two approaches is briefly mentioned in the final chapter.

The term "de Sitter relativity" was coined by Freeman Dyson in a famous paper from the Seventies [1], in which he formulated the main concept of PSR (invariance of physical laws with respect to the de Sitter group, rather than the Poincaré group), and lamented the fact that it had never been seriously explored, as it actually would have deserved. That is why he included dSR among the "missed opportunities" of theoretical physics.

Indeed, when Dyson published his work (1972) both PSR and PGR had been extensively developed by Arcidiacono, starting from Fantappié's original idea of about twenty years earlier [2]. Unfortunately, Arcidiacono published almost exclusively in Italian and mostly in mathematical journals not read by physicists. Fantappié's [2] original work, too, to which we shall return, was published in Italian and has never been translated (to our knowledge) in more widely known languages.

The spread of dSR had thus been hampered by these circumstances; furthermore, there were difficulties of a technical nature. Indeed, dSR locally coincides with the ordinary relativity theory and the only difference from it lies in the data relating to the observation of objects that are very distant in space or events that are very distant in time. The predictions of the two theories differ, therefore, only from a cosmological standpoint; but, owing to the absence of controlled conditions, it is very difficult in this context to conduct crucial tests capable of highlighting unequivocally projective effects. This was probably the most important reason for the substantial lack of interest in this type of approach.

While, as we shall see later, there are considerable problems in ascertaining projective effects by direct experimentation, it is possible today to compare PGR cosmology with ordinary GR cosmology in terms of overall consistency with observational data.

1.2 Historical Origins

Luigi Fantappié was a pure mathematician by profession. He gained his degree at the Pisa *Scuola Normale* with Luigi Bianchi and had later been assistant to Vito Volterra. His rapid academic career as a professor in Italy's most important universities was favoured by his original research on analytical functionals in the complex plane, which earned him international fame. He was at the height of his career when, in 1942, he began to explore the symmetries of physical laws. His research (conducted during Italy's isolation resulting from the Second World War) was especially focused on that which can be defined as an "Erlangen program" for physical science [3].

Basically, Fantappié observed that both the invariance group adopted by Galileo relativity (Galileo group) and that adopted by Einstein relativity (Poincaré group) had the same number of parameters (i.e. 10) and operated in spaces having identical dimensionality (3 spatial dimensions + 1 time dimension). However, while spatial rotations (3 parameters) and boosts (3 parameters) constitute distinct subgroups in the Galileo case, these transformations are combined in the single subgroup of (3 + 1)-dimensional rotations in the Einstein case—the Lorentz group. This unification does not involve spatial translations (3 parameters) and time translations (1 parameter), which remain subgroups that are both mutually distinct and distinct from the Lorentz group. Together, all these transformations form the Poincaré group. Clearly, the unification is made possible by the finite value of the speed of light in vacuum c, which induces the well-known "mixing" of spatial coordinates with the temporal one.

The Galileo group is therefore the limit of the Poincaré group as $c \to \infty$. Can a group with 10 parameters operating on a (3 + 1)-dimensional space exist, more extensive than the Poincaré group, which admits the Poincaré group as an appropriate limit case? Fantappié observes [2] that an n-dimensional space having constant curvature must have a group of isometries with $n(n + 1)/2$ parameters. For $n = 4$ there are 10 parameters, and this result allows to answer the question in the affirmative—the group we are looking for is the de Sitter group associated with a (3 + 1)-dimensional space with curvature radius r, and the Poincaré group is obtained again as the limit as $r \to \infty$.

A second question, therefore, is whether this group can be the limit of some other 10-parameter group on a (3 + 1)-dimensional space. The answer is negative, because the de Sitter group is isomorphous to the 5-dimensional rotations group (O_5) and, as can easily be seen from Cartan's classification, this is a simple group. And a simple group cannot be obtained as the limit case of another simple group. For this reason, Fantappié called the new theory of relativity associated with the de Sitter group "final relativity", a name which has now fallen into disuse. We observe here that the problem addressed by Fantappié was the opposite one to that of group contraction; this problem is much more difficult than the direct problem, which at that time was itself quite difficult. For a comparison between Fantappié's reasoning and contemporary research by Segal and Inönü-Wigner, we remind interested readers to [4].

Once the invariance group of the new relativity was identified, it was necessary to construct the coordinate transformations that generalized the Lorentz-Poincaré transformations (which in turn were extensions of Galileo transformations). It was also necessary to construct the kinematics, dynamics, electromagnetism and thermodynamics of the new theory. This was the task which constituted the life's work of Fantappié's main disciple—Giuseppe Arcidiacono.

Immediately after his degree in physics in Catania in 1951, Arcidiacono had joined Fantappié in Rome, becoming his assistant at the *Istituto Nazionale di Alta Matematica* (INDAM).[1] He continued to work at the INDAM until 1969, when he moved to the University of Perugia, where he remained practically up to his death in 1998.

---

[1]National Institute of Higher Mathematics.

Arcidiacono's initial reasoning was simple: since de Sitter space was curved, it was worthwhile projecting it onto a (3 + 1)-dimensional hyperplane tangent to it at the observation point-event. The new relativity needed to be studied on this flat projection, because it is on it that the observer coordinates remote events. A projective methodology is therefore adopted (from which the alternative name PSR is derived). It is to be noted that the plane representation of de Sitter space-time, called "Castelnuovo space-time", had been used by this author as far back as the early 1930s [5], for the precise purpose of studying kinematics on de Sitter space.

The situation is shown in Fig. 1.1. In 5-dimensional space, de Sitter space appears as a 4-dimensional hyperboloid [6]. A perpendicular to the hyperboloid passing through the projection centre intersects the hyperboloid in the observation point-event. The 4-dimensional hyperplane tangent to the hyperboloid in this point is the Castelnuovo space-time. Its intersection with the hyperboloid is the light-cone having its vertex in the observation point-event. The 5-dimensional light-cone having its origin in the centre of the projection intersects the Castelnuovo space-time in two three-dimensional surfaces which are the two sheets of the Cayley-Klein absolute.

The portion of space-time from which the point-event receives physical signals is thus delimited by the past sheet of its light-cone and by the past sheet of its absolute (de Sitter horizon). Similarly, the portion of space-time in which signals are sent from the observation point-event is delimited by its future light-cone and by the future sheet of the horizon.

The Castelnuovo space-time of point-event O is shown in Fig. 1.2. As can be seen, the chronological distance t_0 of the de Sitter horizon from O is the same for any O, and one has $\pm t_0 = \pm r/c$, where r is the curvature radius of de Sitter space-time. It must be noted that t_0 is a constant that is independent from the observer, and the absolute is changed into itself by the de Sitter group transformations. Therefore, the light-cone must have a variable opening (Fig. 1.3). The

Fig. 1.1 Relationship between de Sitter space and its geodetic projection— Castelnuovo space

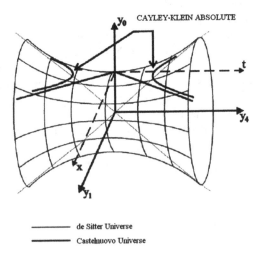

————— de Sitter Universe

————— Castelnuovo Universe

Fig. 1.2 Castelnuovo
space-time and the
observation event-point O,
with its light-cone and its de
Sitter horizon

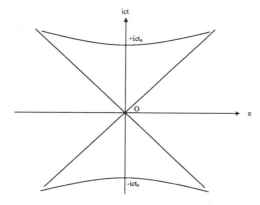

Fig. 1.3 Light-cone with
variable opening

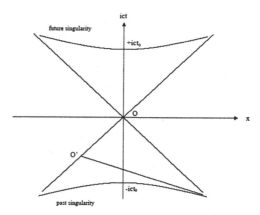

important difference between Minkowski space and Castelnuovo space is the
presence, in the latter, of the de Sitter horizon, which is absent in the former.

We do not detail the Arcidiacono's work here, as it forms the basic structure of
this book. We shall just say that he characterized first PSR (from 1954) then PGR,
starting in 1964. This work did not receive widespread attention, for the reasons
mentioned previously. However, the theory was rediscovered in later years by
various other authors, independently from Arcidiacono.

1.3 Later Developments

The first and important development was the work by the French mathematical
physicists Jean-Marc Lévy-Leblond and Henri Bacry [7]. They again took up the
problem, which had already been treated by Fock [8], of determining the most
general group of frame of reference transformations that leaves free, rectilinear and
uniform motion unchanged. According to their approach, the "kinematic group"

must satisfy three axioms: (1) isotropy of space; (2) time inversion and parity must be automorphisms of the group; (3) the boosts along a given direction must form a non-compact subgroup. From these axioms, the authors derive eleven different kinematics that relate to the de Sitter group and to its space-time, space-velocity, velocity-time, space-velocity-time contractions. This work, which attracted wide-spread interest, also because it was published at a time when attention towards group theory applications was at its highest in the theoretical physicists' community, brought the de Sitter group back to the centre of debate. Up to then, with the sole (and unheeded) exception of Fantappié's and Arcidiacono's work, de Sitter space had always been conceived as an empty (and therefore unphysical), maximally symmetric solution to the Einstein equations with a cosmological constant. The work by Bacry and Lévy-Leblond highlighted, on the other hand, a purely "kinematic" dimension of this group, besides the implications of gravitational theory.

In 1976, Kerner [9] basically finds the same results obtained by Fantappié and Arcidiacono in 1954–1956. However, his reasoning, which is fully original, proceeds in a different way. He reconsiders the fact, demonstrated by Fock, that the most general transformations which preserve free, rectilinear and uniform motion can be expressed as a ratio of linear functions in space-time coordinates. But he confines himself to considering the case in which the denominator is proportional to the scalar product of boost velocity by position. By imposing the customary rule of relativistic composition of velocities, transformations that are essentially 5-dimensional rotations are obtained. In other words, one passes from the general 24-parameter projective group to the 10-parameter O_5 group that is isomorphous to the de Sitter one. Kerner finds, in addition to the speed of light, a new universal constant having the dimensions of a time (or a length).

The direction mapped out by these works is later taken up again, in the first decade of the 21st century, by a Chinese group [10–13] which basically rediscovers (with some interesting extensions) many themes already discussed by Arcidiacono during the 1950s and 1960s of the last century. A Brazilian group retraces the same steps during those same years [14–16]. Starting from this period, it can be said that dSR has come to earn international recognition as a "niche" sector for theoretical research. Themes closely connected with the relation between the de Sitter group, dSR and initial cosmological singularity are now the subject of doctoral theses [17]. More recently, dSR has been extensively described in general textbooks of geometry for physicists [18].

Dyson's admonishment has not gone unheeded.

1.4 PSR as a "Dual" Theory

It is a well-known fact that SR can be interpreted as the effect of a hyperbolic geometry of the space of velocities [19]. This is, indeed, the most natural way of interpreting the existence of a limit velocity and explaining the relativistic law of the composition of velocities.

PSR in this sense is a sort of "double hyperbolic theory" because within it not only the three-dimensional space of velocities relative to the observer is hyperbolic (with a curvature connected to c and a "horizon" that coincides with the light-cone), but so is the three-dimensional space of the distances relative to the observer, i.e. the common space of positions (with a curvature connected to r and a "horizon" constituted by the absolute).

An obvious duality therefore applies between positions and velocities. For example, there is a law on the composition of durations (which will be described in the following chapters) that is formally similar to the law of velocities. This type of duality reminds one of the Poisson brackets duality with respect to the exchange of positions and moments taken with appropriate signs (Born duality). However, no clear connection is known between these two forms of duality.

In SR a precise connection exists between electric and magnetic fields: the latter appear in a reference in which electric charges have non-null velocity with respect to the observer. In other words, the magnetic field appears when the "position" of the sources in the space of the velocities relative to the observer does not coincide with the origin. It is a projective effect caused by this non-coincidence in the hyperbolic space of velocities. If the horizon in this space could be brought to infinity (i.e. if c were increased to infinity) these effects would disappear, and the magnetic field with them. Therefore, by increasing c to infinity in the Maxwell equations, the magnetic field disappears. This is consistent with the fact that the magnetic field is only a projective deformation of the electrical field and does not have charges of its own. There is no evidence of the existence of magnetic monopoles.

In PSR, another situation is added to this one, between the magnetic field and the so-called "C" field, a further component of the electromagnetic field which we shall become acquainted with later. The C field appears in a reference in which the currents that generate the magnetic field are not co-local with the observer. In other words, their position in space of the distances relative to the observer does not coincide with the origin. The C field is a projective effect on the space of positions (space in the customary sense of the term) caused by the presence of the de Sitter horizon. By increasing distance of the horizon to infinity, i.e. considering the limit as r tends to infinity, the C field disappears. Putting $r = \infty$ in the generalized Maxwell equations, the C field disappears; it does not have currents of its own with which to couple. The (electric field, magnetic field) and (magnetic field, C field) pairs are mutually dual.

Let us conclude by observing that the violation of Lorentz invariance induced by PSR holds true at all scales and is independent of the interaction energy, unlike what takes place in Amelino Camelia's "Doubly Special Relativity" (in which the violation increases as energy increases and becomes null at low energies). These two approaches, therefore, must not be confused.

1.5   Experimental Verification

There is no direct evidence at this time that the de Sitter radius r is finite. Testing the finite value of r with direct methods (the only ones which are theoretically capable of providing unambiguous results) is very difficult. A finite radius would certainly show effects similar to a positive cosmological constant:

$$\Lambda \; = \; \frac{3H_0^2}{c^2}\, \Omega_{\Lambda_0} \qquad\qquad (1.1)$$

where symbols have their customary meaning. When the PSR expression $\Lambda = 3/r^2$ and observational values from the Planck Collaboration [20] are substituted in (1.1) one obtains $r \approx 17 \times 10^9$ light years, which can be considered as a lower limit for r. If the cosmological term is entirely due to the projective effects of PSR (as we assume in this volume), this becomes an estimate of r. A similar conclusion was previously drawn from WMAP Collaboration data in [21].

A different direct verification methodology was suggested in the 1970s and 1980s by Arcidiacono himself. He proposed considering the violent expulsion of jets of matter by quasars or active galactic nuclei (AGNs). If the expulsion approximately occurs in the direction of observation, it can reach faster-than-light velocities, with respect to the terrestrial observer. Indeed, as can be seen in Fig. 1.3, the quasar light-cone has, for the terrestrial observer, an opening that is greater than a locally measurable one and which increases with the distance from Earth.

Indeed, cases of faster-than-light components in jets expelled from different quasars have been known ever since the 1970s. In any case, the ballistic model allows clearly to explain these faster-than-light velocities as being purely apparent, without departing from the SR context. If it were possible to measure the two components of the jet velocity (radial and transversal) in a separate and model-independent way, it would be possible to ascertain the true or apparent nature of these velocities and therefore to establish whether any genuine effects of PSR actually do exist. But this goes completely beyond the possibilities of current astrophysics.

Several authors [22–27] tried to estimate experimental higher limits for the discrepancies between PSR and SR predictions, in order to constrain the value of r. It seems, however, that these proposals do not take into account some precautions that are necessary in the practical use of the PSR; cautions motivated by the essential difference between the characters of the inertial frames of reference in these two theories. Recall that the PSR physical spacetime is that of de Sitter, and then only the points that belong to this space are actually point-events. When the de Sitter space is projected onto the plane tangent to it at a point, only that point of the plane belongs to the de Sitter space and is therefore a physical point-event; the other points of the plane are not events. In other words: only the tangential point is a physical event; in particular, the reception (or transmission) of signals by an observer can be identified with this point only. Thus the observer PSR is always placed at the fixed point of the projection, never elsewhere. Other points of the Castelnuovo plane represent the reconstruction, done by the observer, of events actually placed in de Sitter space, to which it is connected through the exchange of signals. The Castelnuovo plane is the *private spacetime* of that observer, while the de Sitter space is the *public spacetime* consisting of all possible observation points-events. These terms were originally introduced by Milne in the context of his Kinematic Relativity and reused by Arcidiacono in the development of the PSR. Obviously, in the ordinary SR limit (i.e., for $r \rightarrow \infty$) the de Sitter space comes to coincide with the tangent plane (which in turn collapses in Minkowski spacetime), and then *every point* of the plane becomes both a point-event that a possible location for the observer associated with a given frame of reference. In other words, public and private spaces coincide. This property is however peculiar to the SR as a limit case, and is not a general feature of the PSR. If we adopt the convention to place, on the tangent plane, the origin of the frame of reference associated with an observer in correspondence with the tangential point, we can say that in PSR the observer is always placed in this origin, while in SR it can be placed in any point of its frame of reference.

In SR it is possible to translate a frame of reference keeping it on the same tangent plane. In PSR, instead, the translation of the frame of reference on the de Sitter space implies a new tangential point for the projection and then the passage to a new tangent plane. The projective representation of the passage from one frame of reference to another (and from one plane to another) is given by the Arcidiacono transformations, which generalize those of Poincaré valid in SR. The Arcidiacono transformations connect the private spaces of two observers (or, more in general, of two point-events) and then describe the projective deformation undergone by a signal that connects these two observers (or points-events). For example, this deformation will contribute to the red shift.

Instead, the properties of an interaction between physical systems that occurs at a certain point-event are purely local aspects which are not influenced by the projective deformation; this applies, for example, to the observable quantities exchanged by the systems (energy, momentum, angular momentum and so on), or interaction parameters (for example, charges). These quantities are measurable only by observers, apparatuses or processes placed at the interaction point-event, i.e., "colocal" with this event. When signals coming from an interaction event are received by a remote observer, the information gained by this observer, through the decoding of these signals, about the physical quantities exchanged in that event is necessarily related to the values of these quantities as defined within the source of the signal, at the instant of its emission. These are inevitably the same values measured by observers which are colocal with that event. Now, the frames of reference colocal with that event are connected by Lorentz transformations (the local limit of the PSR is the SR); therefore, the relations between these quantities will be invariant *with respect to the Lorentz group*, not with respect to Arcidiacono transformations. Of course, the passage from a frame of reference which is colocal with the interaction event to another which is not modifies such relations, making them invariant with respect to the de Sitter group. But these "generalized" relations relate to quantities which are not accessible to any observer and therefore not measurable or, if we like, "counterfactual". These relations are not experimentally testable.

These considerations apply in particular to the dispersion law for a material point, which at first sight seems the obvious "probe" for PSR effects. In fact, the dispersion law valid in PSR is purely formal in nature, as will be explained in more detail in Sect. 4.9. The momentum and energy of a particle measured here and now (or exchanged with other particles in an interaction vertex) are connected together by the usual dispersion law of SR. Thus, any observer, reconstructing the kinematics of an interaction vertex where this particle is involved, will obtain the usual dispersion law of SR. The dispersion law of PSR connects non-measurable quantities, defined in a remote point of Castelnuovo plane that is not accessible to the observer (and that is not really an event), since this latter is confined in the origin of its frame of reference. Consequently, is not possible to constrain r using the dispersion law of PSR or other similar relations.

This, however, is what has been attempted in some papers. For example, in Ref. [27] the impact of high energy cosmic protons with photons of the cosmic microwave background radiation has been studied. In the proton frame of reference the photon has a high energy and can be converted into pions. This effect should reduce the mean free path of cosmic protons above a given threshold energy, making them invisible to a terrestrial observer. It is called GZK suppression [by Greisen-Zatsepin-Kuzmin] but its very existence is controversial. The authors investigate the possibility of a depletion of this efect due to the dispersion law of PSR, without any conclusive result. This conclusion is in accordance with the present discussion, because the relevant dispersion law for protons and photons is in fact that of the SR.

Other papers [22, 23] have been considering the possibility of detecting superluminal movementsin PSR; however, as will be highlighted in Chap. 4, only bodies placed at cosmological distances from the observer can be endowed with superluminal motion with respect to it, and this brings us back to the difficulties already mentioned. Other proposals aim to constrain violations of the equivalence principle in binary star systems [26] or changes in the fine structure constant [24, 25] caused by the PSR. Actually, the PSR has no effect on the local physics and interactions, so these methods can not constrain r.

In a more general perspective, it should also take into account that the difference between PSR and SR is manifested only on the scale of de Sitter radius r, which is certainly cosmological. But on that scale, the PSR is definitely not correct because it represents an *empty* cosmological model. In our opinion the real motivation of the PSR is that it provides an inertial structure which is a suitable foundation for the PGR; the problem of experimental verifications should be posed relatively to the PGR. The situation is similar to that of the Bohr atomic model, which is used to familiarize the student with the concepts and ideas of the quantization; when this is done, one passes to orbitals. The problem of finding an experimental set up able to prove that electrons actually move along elliptical orbits could be ill posed.

1.6   De Sitter Observers, Singularities and Wick Rotations

The view here outlined shows many deep and vital points of contact with the hottest themes in contemporary theoretical physics. There exists unlimited literature, and we will try to sketch out a coarse-grained map of it. The de Sitter metric has been often used, in this regard, in a heuristic way, without specifying the global relativity associated with it. We are convinced that projective methods can suggest unexplored developments of many current approaches.

As it is clear, an "empty" universe will have to await the quantum cosmology developments to fuel the interest of the physicists. In the context of the FRW-type cosmologies the existence of a pre-inflationary phase characterized by a de Sitter metric is generally admitted, a line of thought started by the works of J. Hartle, S. Hawking and A. Vilenkin [28–31]. These works—and the long series they gave rise to (see [29] for a review)—contextually raised the hard problem to define a Quantum Field Theory (QFT) on de Sitter universe (excellent review articles are [32–36]). Comparing the cosmological research and QFT on the de Sitter space the global impression is that we are facing a paradox that, brutally, can be thus described: building a QFT on the de Sitter space appears extraordinarily easy thanks to its maximal symmetry, in particular for what concerns the possibilities of a "finite" Hilbert-Fock space as well as a "natural selection" of fluctuations. However, these results do not seem fully in agreement with the inflation.

This seems to suggest the necessity of radically new approaches. This is the objective of the holographic principle, introduced from G. t'Hooft and L. Susskind in the first '90s [37–40]. The idea was first formulated in the context of black holes thermodynamics; it substantially affirms that the amount of information contained in a physical system is proportional to its area, instead of to its volume. In practice, the source of the dynamics is the information coded on the system surface and "projected" in the system volume. The mechanism able to connect organically information (emergence of space and time) and dynamics (observables in space and time) is still lacking. The here outlined projective theory can be a suggestion in this regards. A particularly important result was obtained by J. Maldacena with the famous AdS/CFT conjecture, that postulates a duality relation between two classes of apparently very different theories: the conformal supersymmetric Yang-Mills of order $N = 4$ and the theories of strings on an anti-de Sitter five-dimensional universe. A similar result was later obtained on dS by A. Strominger [41–44]. We recall that dS space is associated with expansion in real time, positive cosmological constant, negative curvature while AdS space is associated with contraction in real time, negative cosmological constant and positive curvature.

Since transformations from dS to AdS and reverse exist (these spaces correspond, in imaginary time, to the same hypersphere S^4, see Fig. 1.4), these results make rather problematic the interpretation of the holographic principle, and in any case really different from the t'Hooft-Susskind one. In fact a N-dimensional bulk universe and its $(N - 1)$-dimensional projection seem to be involved here (see [45] for a recent model).

Fig. 1.4 AdS space in real
and imaginary time

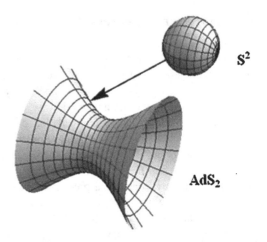

$$\mathbf{S^2}$$

$$\mathbf{AdS_2}$$

Another interesting, even though marginal today, line of research regards the models of elementary particles based on the de Sitter metric. While effective field theories are based on the procedure of renormalization which is well definite nowadays, non-perturbative approaches are still waiting for new ideas about quantum gravity [46, 47].

In the following we outline some possible connections between these ideas and the projective approach, according to [48].

From a quantum viewpoint the S^4 interesting aspect is that it is at imaginary cyclic time and without singularities. It means that it is impossible to define on de Sitter a global temporal coordinate. So it has an istanton feature, individuated by its Euler topological number which is 2 [49]. This leads to a series of formal analogies both with black holes' quantum physics and the theoretical proposals for the "cure" for singularities.

Let us consider the De Sitter-Castelnuovo metric in real time:

$$ds^2 = -\left(1 - \frac{H^2 r^2}{c^2}\right) dt^2 + \left(1 - \frac{H^2 r^2}{c^2}\right) dr^2 + r^2 d\Omega^2 \qquad (1.2)$$

where $d\Omega^2 = d\theta^2 + \sin^2\theta d\varphi^2$ in polar coordinates.

As we have seen in PSR, the singularity in $r = c/H$ becomes an horizon of events for any observer when it passes to the Euclidean metric with $\tau \rightarrow -it$:

$$ds^2 = d\tau^2 + \frac{1}{H^2} \cos H\tau \left(dr^2 + \sin^2 r d\Omega^2\right) \qquad (1.3)$$

with a close analogy with the Schwarzschild solution's case. The τ period is $\beta = 2\pi/H$; for the observers in De Sitter it implies the possibility to define a temperature, an entropy and an area of the horizon, respectively given by:

$$T_b = \frac{H}{2\pi} = \beta^{-1}; \quad S = \frac{\pi}{H^2} = \frac{\beta^2}{4\pi}; \quad A = \frac{4\pi}{H^2} = \frac{\beta^2}{\pi}. \tag{1.4}$$

From (1.4) we get the following fundamental outcome:

$$S = \frac{1}{4} A \tag{1.5}$$

which is the well-known expression of the t'Hooft-Susskind-Bekenstein holographic prin-ciple. The (1.5) connects the non-existence of a global temporal coordinate with the information accessible to any observer in the de Sitter model. In this way we obtain a deep physical explanation for applying the Weyl principle in the de Sitter Universe, and sum up that in cosmology, as well as in QM, a physical system cannot be fully specified without defining an observer [50]. G. Arcidiacono stated that the hyper-spherical Universe is like *a book written with seven seals* (Apocalypse, 6–11), and consequently two operations are necessary to investigate its physics: (1) inverse Wick rotation and (2) Beltrami-Castelnuovo representation. That's the way we can completely define a relativity in de Sitter.

The association of imaginary time with temperature gets a remarkable physical significance which implies some considerations on the statistical partition function [51]. For our aims it will be sufficient to say that such temperature is linked to the relation (1.5), i.e. to the information that an observer spent within his area of events. Which thing has patent implications from the dynamical viewpoint, because it is the same as to state that, as well as in Schwarzschild black hole' s case, the De Sitter space and the quantum field defined on it behave as if they were immersed in background fluctuations. The transition amplitude from a configuration of a generic field in $t_2 - t_1 = dt$ time will be given by the e^{-iHdt} matrix element which acts as a $U(1)$ group transformationof the $U(1)_{space} \leftrightarrow U(1)_{time}$. It means that a transition amplitude on S^4 will appear to an observer as the $R(t)$ scale factor's variation with H variation rate.

It makes possible to link the hyper-spherical description with the Big-Bang evolutionary scenario and to get rid of the thermodinamic ambiguities which characterize its "beginning" and "ending" notions. The last ones have to be re-interpretated as purely quantum dynamics of the matter-fields on the hyper-sphere free of singularities.

The Hartle-Hawking proposal of "no-boundary" condition removes the initial singularity and allows to calculate the Universe wave function. In fact, it is possible —as in the usual QFT—to calculate the path integrals by using a Wick rotation as "Euclidization" procedure. In such way also the essential characteristics of the inflationary hypotheses are englobed. The derived formalism is similar to that used in the ordinary QM for the tunnel effect, an analogy which should explain the physics at its bottom [52–54].

The group extension method provides this procedure with a solid foundation, because the de Sitter space, maximally symmetric and simply connected, is univocally individuated by the group structure, and consequently is directly linked to the space-time homogeneity and isotropy principle with respect to physical laws. The original Hartle-Hawking formulation operates a mix of topologies hardly justified both on the formal level and the conceptual one. The "no-boundary" condition is only valid if we work with imaginary time, and the theory does not contain a strict logical procedure to explain the passage to real time. This corresponds to a quite vague attempt to conciliate an hyper-spherical description at imaginary time with an evolutive one at real time according to the traditional Big-Bang scenario. In fact, it has been observed that the Hartle-Hawking condition is the same as to substitute a singularity with a "nebulosity".

The spontaneous proposal, at this point, is considering the Hartle-Hawking conditions on primordial space-time as a consequence of a global charaterization of the hyper-sphere and directly developing quantum physics on S^4. Which thing does not contradict the quantum mechanics formulation and its fundamental spirit, which is to say the Feynman path integrals. In other words, quantum mechanics has not to be applied to cosmology for the Universe smallness at its beginning, but because each physical system—without exception—gets quantum histories with amplitude interferences. The "by nothing creation" means that we cannot "look inside" an istanton (hyper-spherical space), but we have to recourse to an "evolutionary" description which separates space from time. The projective methods tell us how to do it.

An analogous problem—to some extent—is that of the Weyl Tensor Hypothesis. Roger Penrose has suggested a condition on the initial singularity that, within the GR, ties entropy and gravity and makes a time arrow emerge [55, 56]. It is known that the W_{ABCD} Weyl conformal tensor describes the freedom degrees of the gravitational field. The Penrose Hypothesis is that $W_{ABCD} = 0$ in the Big-Bang, while $W_{ABCD} = 1$ in the Big-Crunch. The physical reason is that in the Universe's initial state we have an highly uniform matter distribution at low entropy (entalpic order), while in Big-Crunch, just like a black hole, we have an high entropy situation. This differentiates the two singularities and provides a time arrow. In an hyper-spherical Universe there is no "beginning" and "ending", but only quantum transitions. Consequently, the Penrose Hypothesis can only be implemented in terms of projective representation within the ambit of PGR.

Finally, we can take into consideration the possibility to build a Quantum Field Theory on S^4. A QFT, for T tending towards zero, is a limit case of a theory describing some physical fields interacting with an external environment at T temperature. Without this external environment we could not speak of decoherence, could not introduce concepts such as like dissipation, chaos, noise and, obviously, the possibility to describe phase transitions would vanish too. Therefore, it is of paramount importance to write a QFT on de Sitter background metric and then studying it in projective representation. If we admit decoherence processes on S^4, it is possible to interpret the Weyl Principle in a new, strong form: the "classical" and observable Universes are the ones where it can be operated a description at real time.

In conclusion, it is possible to delineate an alternative, but not incompatible with tra-ditional cosmology scenario. The Universe is the quantum configuration of the quantum fields on S^4. Thus developing a Quantum Cosmology coincides with developing a Quantum Field Theory on a space free of singularities. The Big-Bang is a by vacuum nucleation in an hyper-spherical background at imaginary time, and so the concepts of "beginning", "expansion" and "ending" belong to the space-time foreground and gain their meaning only by means of a suitable representation which defines a family of cosmological observers.

1.7 De Sitter Relativity in Current Research

In this book the de Sitter radius r is introduced as a new universal constant (on the same foot of the speed limit c), and so the cosmological constant Λ it induces. Passing from PSR to PGR in order to take gravitation into account (Chap. 6), the constancy of r is retained. However, it is necessary to emphasize the existence of different approaches to the one presented in this volume, according to which the generalization of the PSR, performed in order to include gravitation, does not preserve the constancy of Λ. We limit ourselves here to mentioning these approaches, referring the interested reader to the relevant literature, because they are significantly different from our own [11–16, 57–59].

The reason to adopt Λ as a variable quantity consists in the possibility of being able to embrace, with a single theory: the situation in which r was in the order of the Planck length (situation that is assumed could have occurred to the big bang), the next inflationary phase, when Λ decreased roughly to current levels and finally the post-inflationary phase, where Λ reappears as the acceleration of the cosmic expansion. In addition to this possibility of a unified description of the entire history of the Universe, certainly tempting, a second reason for considering Λ as a variable is the physics of extremely high energies. Indeed, also in the present Universe a process of interaction between elementary particles can induce a notable bending of the spacetime, near the point where the interaction occurs. This is possible if the exchanged energy is comparable to that of Planck; in this case a de Sitter microspace could in principle be generated, with a decreasing radius when the exchanged energy increases. In extreme conditions of this type, the spacetime geometry is often described by means of continuous deformations of the Lorentz group which, however, have the unfortunate drawback of also deforming the definition of relevant observables, in addition to causal problems induced by the deformation of the light cone. The PSR would not present these drawbacks, since its dispersion laws remain valid at all scales, and the light cone remains unchanged (unless a scaling) when r changes. In the extreme limit of a vanishing de Sitter radius, the light cone degenerates becoming invariant for conformal transformations only; an extreme condition in which the Universe would consist of only dark energy. Such an approach widely exploits the decomposition of the de Sitter translations in a "conventional" SR part and a contribution derived by conformal transformations [60].

In [59] the generalization of the PSR along these lines leads to a pair of gravitational equations, respectively containing the energy-momentum tensor and the
conformal current of matter. The first equation is in fact the well-known gravitational equation of GR without cosmological term; the second equation describes a
dependence of the cosmological constant on the Universe matter content. This
second equation is the real novelty of the approach; for a universe of pure powder it
leads to the expression:

$$\Lambda = \frac{4\pi G}{c^2} \mu \qquad (1.6)$$

where μc^2 is the energy density, G is the gravitational constant and c is the speed
limit. Substituting the density value, derived by observations, for the present
Universe one obtains $\Lambda \approx 10^{-56}$ cm^{-2}, which agrees in order of magnitude with
the observations. It is remarkable that starting from our own approach the following
relations are derived [61]:

$$\mu = \frac{\tau^{-2}}{6\pi G} ; \quad \Lambda = \frac{4}{3r^2} \qquad (1.7)$$

where τ is the cosmic time, that for the present Universe ($\tau \approx r/c$) return the
previous equation unless a factor two. This could mean that observations limited to
the nearby (recent) Universe are not sufficient to distinguish the two approaches.

In their construction the authors of [59] assume the validity of the relation
$\Lambda = 3/r^2$ by borrowing it from the GR and use it to define Λ. Of course, here r is
no longer a *global* property of space as in the case of the original PSR, where this
parameter, independent on position and time, was related to the scalar curvature of
de Sitter Universe. The parameter r now defines the radius of the de Sitter space
osculating the spacetime at a determined point-event, and as such it is a function of
that point-event. The subsequent application of the cosmological principle limits
this dependency to only cosmic time.

There have also been attempts to introduce the cosmological principle, and then
the cosmic time, directly in the formal structure of the PSR, in order to obtain a full
representation of the inertial background. In [62] the transformation that leads from
the Beltrami projective time (the time coordinate of an observer on the plane of
Castelnuovo) to the observer's time marked by its local clock is considered.
Obviously, the local time is unbounded while the projective time is bounded by the
two sheets of de Sitter horizon. The time marked by the clock colocal with the
observer is identified by the authors with cosmic time, but this identification does
not seem entirely convincing. As a matter of fact, the PSR Universe is empty of
matter and therefore devoid of physical variables (temperature, pressure and so on)
on the values of which distinct fundamental observers may synchronize their
clocks. There is no cosmic fluid in the PSR, and therefore may not even exist a
cosmic time. The de Sitter universe is empty everywhere and always, and all the
fundamental observers will see it, always and everywhere, as empty. In a sense

there is here a kind of "perfect" cosmological principle, which does not allow the possibility to define a cosmic time. We will see in Chap. 6 as the result proposed by the authors of [62] can be recovered in the PGR.

These notes are only introductory to an elementary study of theory. The interested reader can find in [63] a detailed discussion of the current status of the research in this field.

1.8 De Sitter Space and Field Theory

Even though this book is dedicated to the de Sitter Relativity, it is necessary at least make mention of the major implications of de Sitter geometry in Quantum Field Theories (QFT).

How it is well known since a long time, the de Sitter space is an ideal background for the construction of field theories, thanks to its peculiar symmetry conditions. Particularly remarkable aspects are those that fuel the hopes in the possibility of constructing non-perturbative theories [64, 65]. These research lines intersect with the vast literature raised by the Hartle-Hawking cosmological solution, with interesting implications about the balance of positive/negative modes, quantum fluctuations and spacetime dimensionality [66–69]. It is difficult today to make a clear summary of this researches. In general, it is possible to consider as an important discriminant the role assigned to the cosmological "constant" and its physical interpretation: is it a dynamical variable in the evolution of universe or rather a structural and "fixed" geometrical property? These are two different physical interpretations whose consequences are relevant with regard to quantum theory [70].

These issues are related to the fundamental Maldacena conjecture, which establishes a duality relationship between AdS geometry and Conformal Quantum Field Theory (CFT). It is still a conjecture in search of a theory: strings are the favorite candidate, but there are other approaches [71]. The Maldacena work is the origin of the holographic theories, in which a correspondence exists between a given spatiotemporal structure of dimensionality $n \geq 5$ and the activity of physical objects in 4d. However, as Lev writes: "The experimental fact that $\Lambda > 0$ might be an indication that for some reasons nature prefers dS invariance vs AdS invariance ($\Lambda < 0$) and Poincaré invariance ($\Lambda = 0$)". In fact it is possible to propose a dS/CFT conjecture [72] in accordance with the requests of a projective holography, in search of a more intimate connection between quantum theory and spacetime [73, 74].

References

1. Dyson, F.J.: Missed opportunities. Bull. Am. Math. Soc. **78**, 635–652 (1972)
2. Fantappié, L.: Rend. Accad. Lincei XVII, fasc. 5 (1954)

3. Fantappié, L.: Sui fondamenti gruppali della fisica (posthumous). Collectanea Math. XI, fasc. 2 (1959)
4. Ciccoli, N.: Fantappiè's "final relativity" and deformations of Lie algebras. Arch. Hist. Exact Sci. **69**(3), 311–326 (2015)
5. Castelnuovo G.: Rend. Accad. Lincei XII, 263 (1930)
6. Moschella U.: The de Sitter and anti-de Sitter sightseeing tour. In: Damour, T., Darrigol, O., Duplantier, B., Rivesseau, V. (eds.) Einstein 1905–2005, Progress in Mathematical Physics 47. Birkhauser, Basel (2006)
7. Bacry, H., Lévy-Leblond, J.: Possible kinematics. J. Math. Phys. **9**(10), 1605 (1968)
8. Fock, V.A.: The Theory of Space, Time and Gravitation. Pergamon Press, Oxford (1964)
9. Kerner, H.E.: Proc. Natl. Acad. Sci. U. S. A. **73**, 1418–1421 (1976)
10. Guo, H.-Y.: On principle of inertia in closed Universe. Phys. Lett. **B653**, 88–94 (2007)
11. Guo, H.-Y.: Special relativity and theory of gravity via maximum symmetry and localization. Sci. China A **51**(4), 568–603 (2008)
12. Guo, H.-Y., Huang, C.-G., Xu, Z., Zhou, B.: On beltrami model of de Sitter Spacetime. Mod. Phys. Lett. A **19**, 1701–1710 (2004)
13. Guo, H.-Y., Zhou, B., Tian, Y., Xu, Z.: The triality of conformal extensions of three kinds of special relativity. Phys. Rev. D **75**, 026006 (2007)
14. Aldrovandi, R., Beltrán Almeida, J.P., Pereira J.G.: Some implications of the cosmological constant to fundamental physics. In: Cosmology and Gravitation, XIIth Brazilian School of Cosmology and Gravitation. AIP Conference Proceedings 910, pp. 381–395 (2007)
15. Aldrovandi, R., Beltrán Almeida, J.P., Mayor, C.S.O., Pereira, J.G.: Lorentz transformations in de Sitter relativity. gr-qc/0709.3947
16. Aldrovandi, R., Beltrán Almeida, J.P., Pereira J.G.: De Sitter special relativity. Class. Quantum Grav. **24**(6), 1385–1404 (2007)
17. Janzen D.: A solution to the cosmological problem of relativity theory. Ph.D. Thesis, University of Saskatchewan, March 2012
18. Aldrovandi, R., Pereira, J.G.: An Introduction to Geometrical Physics, 2nd edn. World Scientific, Singapore (2017)
19. Barrett, J.F.: The hyperbolic theory of special relativity. arXiv:1102.0462[physics.gen-ph]
20. Planck Collaboration. arXiv:1502.01589
21. Adler, R.J., Overduin, J.M.: The nearly flat Universe. Gen. Rel. Grav. **37**, 1491–1503 (2005)
22. Yan, M.-L., Xiao, N.-C., Huang, W., Hu, S.: Superluminal neutrinos from special relativity with de Sitter space-time symmetry. Mod. Phys. Lett. A **27**, 1250076 (2012)
23. Yan, M.-L., Hu, S., Huang, W., Xiao, N.-C.: On determination of the geometric cosmological constant from the OPERA experiment of superluminal neutrinos. Mod. Phys. Lett. A **27**, 1250041 (2012)
24. Yan, M.-L.: Evidence for special relativity with de Sitter space-time symmetry. Chin. Phys. C. **35**, 228–232 (2011)
25. Feng, S.-S., Yan, M.-L.: Implication of spatial and temporal variations of the fine-structure constant. Int. J. Theor. Phys. **55**(2), 1049–1083 (2016)
26. Tretyakova, D.A.: Seeking for the observational manifestation of de Sitter relativity. Grav. Cosm. **22**, 339–344 (2016)
27. Chang, Z., Chen, S.-X., Huang, C.-G.: Absence of GZK Cutoff and test of de Sitter invariant special relativity. Chin. Phys. Lett. **22**, 791–794 (2005)
28. Bojowald, M.: Quantum cosmology: a review. Rep. Prog. Phys. **78**, 2 (2015)
29. Bojowald, M.: Quantum Cosmology: A Fundamental Description of the Universe. Springer (2011)
30. Hartle, J.B., Hawking, S.W.: Wave function of the Universe. Phys. Rev. D **28**, 12 (1983)
31. Vilenkin, A.: Creation of Universes from nothing. Phys. Lett. **117b**, 1/2 (1982)
32. Gazeau, J.P., Lachieze Rey, M.: Quantum field theory in de Sitter space: a survey of recent approaches. arXiv:hep-th/0610296 (2006)
33. Giddins, S.B., Marolf, D.: A global picture of quantum de Sitter space. Phys. Rev. D **76**, 064023 (2007)

34. Takook, M.V.: Quantum field theory in de Sitter Universe: ambient space formalism. arXiv: 1403.1204 [gr-qc] (2014)
35. Parsamehr, S., Enayati, M., Takook, M.V.: Super-gauge Field in de Sitter Universe. arXiv: 1504.00453 [gr-qc] (2015)
36. Albrecht, A., Holman, R., Richard, B.J.: Equilibration of a quantum field in de Sitter space-time. Phys. Rev. D **91**, 043517 (2015)
37. t'Hooft, G.: Dimensional Reduction in Quantum Gravity. arXiv:gr-qc/9310026 (1993)
38. Susskind, L.: World as a hologram. J. Math. Phys. **36**, 6377–6396 (1995)
39. Bousso, R.: The holographic principle. Rev. Mod. Phys. **74**(3), 825–874 (2002)
40. Page, D.N.: Susskind's challenge to the Hartle–Hawking no-boundary proposal and possible resolutions. JCAP 0701 (2007)
41. Maldacena, J.: The large-N limit of superconformal field theories and supergravity. Int. Jour. Theor. Phys. **38**(4), 1113–1133 (1999)
42. Braga, N.R.F.: Quantum fields in Anti-de Sitter space and the Maldacena conjecture. Braz. J. Phys. **32**(4), 880–883 (2002)
43. Strominger, A.: The dS/CFT correspondence. JHEP **10**, 034 (2001)
44. Witten, E.: Anti De Sitter space and holography. Adv. Theor. Math. Phys. **2**, 253–291 (1998)
45. Pourhasan, R., Afshordi, N., Mann, R.B.: Out of the white hole: a holographic origin for the big bang. JCAP 1404 (2014)
46. Smrz, P.K.: Geometrical models of elementary particles. Aust. J. Phys. **48**(6), 1045–1054 (1995)
47. Guo, H., Huang, C., Tian, Y., Xu, Z., Zhou, B.: Snyder's quantized space-time and de Sitter special relativity. Front. Phys. China **2**(3), 358–363 (2007)
48. Licata, I.: Universe without singularities. A group approach to De Sitter cosmology. Electr. J. Theor. Phys. **3**(10), 211–224 (2006)
49. Rajaraman, R.: Solitons and Istantons 15. North-Holland Publ, New York (1982)
50. Rugh, S.E., Zinkernagel, H.: Weyl's Principle, Cosmic Time and Quantum Fundamentalism. arXiv:1006.5848 [gr-qc] (2010)
51. Hawking, S.W.: Particle creation by black holes. Comm. Math. Phys. **43**(3), 199–220 (1975)
52. Frolov, V.P., Markov, M.A., Mukhanov, V.F.: Black holes as possible sources of closed and semiclosed worlds. Phys. Rev. D **41**, 2 (1990)
53. Borde, A., Guth, A., Vilenkin, A.: Inflationary spacetimes are not past-complete. Phys. Rev. Lett. **90**, 151301 (2003)
54. Hawking, S.W., Moss, I.G.: Fluctuations in the very early universe. Nucl. Phys. B **224**, 180 (1983)
55. Penrose, R.: Singularities and time-asymmetry. In: Hawking, S.W., Israel, W. (eds.) General Relativity: An Einstein Centenary Survey, pp. 581–638. Cambridge University Press (1979)
56. Tod, P.: Penrose's Weyl curvature hypothesis and conformally-cyclic cosmology. J. Phys. Conf. Ser. 229 (2010)
57. Aldrovandi, R., Pereira, J.G.: De Sitter relativity: a new road to quantum gravity? Found. Phys. **39**, 1–9 (2009)
58. Aldrovandi, R., Pereira, J.G.: Is physics asking for a new kinematics? Int. J. Mod. Phys. D **17**, 2485 (2008)
59. Beltrán Almeida, J.P., Mayor, C.S.O., Pereira J.G.: De Sitter relativity: a natural scenario for an evolving Λ. Grav. Cosmology **18**(3), 181–187 (2012)
60. Aldrovandi, R., Beltrán Almeida, J.P., Pereira, J.G.: Some implications of the cosmological constant to fundamental physics. AIP Conf. Proc. **910**, 381 (2007)
61. Chiatti, L.: De Sitter relativity and cosmological principle. TOAAJ **4**, 27–37 (2011)
62. Guo, H.-Y., Huang, C.-G., Xu, Z., Zhou, B.: On special relativity with cosmological constant. Phys. Lett. A **331**, 1–7 (2004)
63. Cacciatori, S., Gorini, V., Kamenshchik, A.: Special relativity in the 21st century. Ann. Der Physik **17**, 728–768 (2008)
64. Gazeau, J.P., Lachiéze-Rey, M.: Quantum field theory in de Sitter space: a survey of recent approaches. arXiv:hep-th/0610296 (2006)

65. Giddings, S.B., Marolf, D.: A Global picture of quantum de Sitter space. Phys. Rev. D **76**, 064023 (2007)
66. Castro, A., Maloney, A.: The wave function of quantum de Sitter. arXiv:1209.5757 [hep-th]
67. Mbarek, S., Paranjape, M.B.: Negative mass bubbles in de Sitter space-time. Phys. Rev. D **90**, 101502(R) (2014)
68. Verdaguer, E.: Gravitational fluctuations in de Sitter cosmology. J. Phys. Conf. Ser. **314**(1), 012008 (2011)
69. Momen, A., Rahman, R.: Spacetime Dimensionality from de Sitter Entropy. arXiv:1106.4548 [hep-th]
70. Lev, F.M.: Positive cosmological constant and quantum theory. Symmetry **2**(4), 1945–1980 (2010)
71. Dolce, D.: Classical geometry to quantum behavior correspondence in a virtual extra dimension. Ann. Phys. **327**(9), 2354–2387 (2012)
72. Vistarini, T.: Holographic space and time: Emergent in what sense?. Stud. Hist. Phil. Sci, B (first on line (2016)
73. Chiatti, L.: Choosing the right relativity for QFT. In: Licata, I., Sakaji, A. (eds.) Vision of Oneness, pp. 365–398. Aracne Editrice, Roma. arXiv:0902.1293 [physics.gen-ph] (2011)
74. Licata, I.: In and out of the screen. On some new considerations about localization and delocalization in Archaic Theory. In: Licata, I. (ed.) Beyond Peaceful Coexistence. The Emergence of Space, Time and Quantum, pp. 559–577. Imperial College Press (2016)

Chapter 2
Steps Towards the De Sitter Relativity

It is here introduced the *Erlangen program for Physics* by G. Arcidiacono, where the three relativities (Galilei, Poincaré and de Sitter) are a groupal "matryoshka", i.e., three description levels of physical phenomena. The Arcidiacono transformations for the de Sitter group and the cosmological consequences are also introduced.

2.1 Some Recalls About Lie Groups[1]

Symmetry is one of the most fundamental properties of nature. The branch of mathematics dealing with symmetry is the group theory. Let us recall that a group is a set G endowed with an associative operation denoted ab for group elements $a, b \in G$. The group also contains a unique identity element, denoted u, and each group element a has an inverse a^{-1} satisfying $aa^{-1} = a^{-1}a = u$. Lie groups lie at the intersection of two fundamental fields of mathematics: algebra and geometry. A Lie group is first of all a group, and secondly it is a differentiable manifold. Differentiable manifolds are the basic objects in differential geometry and they generalize to higher dimensions the curves and surfaces. A manifold is a topological space which resembles Euclidean space locally. A Lie group is a group which is also a differentiable manifold, with the property that the group operations are compatible with the differential structure. That is, the applications

(a)
$$G \times G \to G : (a, b) \to ab \tag{2.1}$$

(b)
$$G \to G : a \to a^{-1} \tag{2.2}$$

---

[1]For a more extended treatment of Lie algebras and groups, the reader is referred to classical literature [1–7].

© The Author(s) 2017
I. Licata et al., *De Sitter Projective Relativity*, SpringerBriefs in Physics,
DOI 10.1007/978-3-319-52271-5_2

are differentiable. Informally, a Lie group is a group of continuous symmetries. A Lie algebra is a vector space V over some field K together with a binary operation $[\cdot, \cdot]: V \times V \to V$ called the Lie bracket, which satisfies the following axioms

(a) Bilinearity:

$$[a + b, c] = [a, c] + [b, c], \tag{2.3}$$

$$[a, b + c] = [a, b] + [a, c], \tag{2.4}$$

$$[\lambda a, b] = [a, \lambda b] = \lambda[a, b], \tag{2.5}$$

for all $a, b \in V$ and for all $\lambda \in K$.

(b) Anticommutativity:

$$[a, b] = -[b, a] \text{ for all } a, b \in V. \tag{2.6}$$

(c) The Jacobi identity:

$$[[a, b], c] + [[b, c], a] + [[c, a], b] = 0 \text{ for all } a, b, c \in V. \tag{2.7}$$

From the second axiom we deduce

$$[a, b] + [b, a] = 0 \Rightarrow [a, a] + [a, a] = 2[a, a] = 0 \Rightarrow [a, a] = 0 \tag{2.8}$$

for all $a \in V$. For any associative algebra $(A, *)$ one can construct a Lie algebra where the Lie bracket is defined by commutator $[a, b] = a * b - b * a$.

If $\{e_1, \ldots, e_n\}$ is a basis of V we can introduce the so-called structure constants defined by $[e_i, e_j] = c_{kij}e_k$, and these constants are useful to calculate the commutator of any pair $a, b \in V$. In fact $[a, b] = a_i b_j [e_i, e_j] = a_i b_j c_{kij} e_k$.

Any vector space V endowed with the identically zero Lie bracket becomes a Lie algebra. Such Lie algebras are called Abelian. The three-dimensional Euclidean space R^3 with the Lie bracket given by the cross product of vectors becomes a three-dimensional Lie algebra. After these premises, we analyze some groups of transformation which are particularly important in the study of physics. In an Euclidean n-dimensional space, the linear and homogeneous transformations

$$x_i' = \sum_k a_{ik} x_k \tag{2.9}$$

that leave invariable the quadratic form $x_1^2 + \cdots + x_n^2$ are said to be orthogonal transformations; they form the orthogonal group O_n. The matrix $[a_{ik}]$ is orthogonal, that is its inverse is equal to its transpose. Moreover, the determinant of $[a_{ik}]$ is ± 1, and the orthogonal transformations whose matrices have determinant equal to 1 form a subgroup called special orthogonal group $SO(n)$ which is the group of all rotations about the origin of n-dimensional Euclidean space. The numbers of parameters of the orthogonal and special orthogonal group is $n(n-1)/2$. The orthogonal group is a subgroup of the general linear group GL_n with determinant of $[a_{ik}] \neq 0$ with n^2 parameters. We also quote the special linear group SL_n, the subgroup of GL_n consisting of matrices with determinant equal to 1. Even relevant is the unitary group $U(n)$ of dimension n^2, consisting of $n \times n$ unitary complex matrices, which is a subgroup of the general linear group. Finally we quote the special unitary group denoted $SU(n)$ consisting of all $n \times n$ unitary complex matrices with determinant 1, whose dimension is $n^2 - 1$.

The mathematician Sophus Lie had the idea to consider, in the study of continuous groups of transformations, the infinitesimal elements of group, that is the elements that are in a neighborhood of the identity element. The properties of these infinitesimal elements characterize the properties of the group. If we consider a group of transformations with one parameter t

$$x_i' = x_i(x_1, \dots .x_n, t) \tag{2.10}$$

and we introduce the infinitesimal operator X of the group, it is possible to show that

$$x_i' = x_i + tXx_i + \frac{1}{2}t^2 X^2 x_i + \dots = e^{tX} x_i. \tag{2.11}$$

When the parameter t is considered infinitely small, we obtain the infinitesimal transformations

$$x_i' = x_i + tXx_i. \tag{2.12}$$

For example, the infinitesimal operator of rotations group on a plan is $X = x_1 \partial_2 - x_2 \partial_1$, and we have the infinitesimal transformations

$$x_1' = x_1 - tx_2 \tag{2.13}$$

$$x_2' = x_2 - tx_1 \tag{2.14}$$

Assigned a continuous group with r parameters, it is possible to calculate its infinitesimal operators and its infinitesimal transformations. Viceversa, assigning r infinitesimal operators $X_1, \dots X_r$ and introducing the Lie bracket $[X_i, X_k] = X_i X_k - X_k X_i$ we have r infinitesimal operators, among them independent, that can produce a group with r parameters if and only if $[X_i, X_k] = \sum_s c_{iks} X_s$.

Two groups with the same structure constants are isomorphic in the neighborhood of the identity element. Cartan has given a complete classification of the simple groups, and he has found that there are four infinite families.

(1) A_n groups; a model of these groups is $SU(n+1)$
(2) B_n groups; a model of these groups is $SO(2n+1)$

(3) C_n groups; a model of these groups is the symplectic group $S_p(2n)$
(4) D_n groups; a model of these groups is $SO(2n)$.

Then there are other five possible simple groups and these are the so-called exceptional cases G_2, F_4, E_6, E_7 and E_8. These cases are said exceptional because they do not fall into infinite series of groups of increasing dimension. G_2 has 14 dimensions, F_4 has 52 dimensions, E_6 has 78 dimensions, E_7 has 133 dimensions and E_8 248 dimensions. The E_8 algebra is the largest and most complicated of these exceptional cases and is often the last case of various theorems to be proved. E_8 is a very beautiful group; in fact it is the symmetry group of a particular 57-dimensional object.

2.2 Lie Groups of Spacetime

The spacetime of Newtonian physics is a four-dimensional manifold endowed with the following geometrical structures:

(1) Absolute time
(2) Absolute space
(3) Absolute spatial distances
(4) Absolute temporal distances.

The spacetime of special relativity, instead, is a four-dimensional manifold endowed with the following geometrical structures:

(1) Relative time
(2) Relative space
(3) Absolute Space-Time distances.

Let us remember that Galileo group is the main group of classical physics and is formed by the composition of the following transformations.

(a) Spatial rotations characterized by three parameters

$$x'_\mu = a_{\mu\nu} x_\nu \tag{2.15}$$

where $[a_{\mu\nu}]$ is a 3×3 orthogonal matrix whose determinant is $+1$.

(b) Inertial motions (boosts) characterized by the three components of velocity

$$x'_\mu = x_\mu + v_\mu t \tag{2.16}$$

(c) Spatial translations characterized by three parameters

$$x'_\mu = x_\mu + a_\mu \tag{2.17}$$

(d) Temporal translations characterized by only one parameter

$$x'_\mu = x_\mu, t' = t + t_0 \tag{2.18}$$

The global Galileo group dimension is 10. This group appears in the formulation of Galileo well know relativity principle. Moving on to relativistic physics, spatial rotations and boosts are blended in a unique operation: the rotations of a pseudo-Euclidean space (Minkowski spacetime) M_4, characterized by 6 parameters,

$$x'_i = a_{ik}x_k \tag{2.19}$$

where $x_4 = ict$. These transformations, called Lorentz special transformations, form Lorentz proper group and, after joining reflections, the Lorentz extended group. Then we need to add the translations of M_4

$$x'_i = x_i + a_i \tag{2.20}$$

characterized by 4 parameters, which include spatial and temporal translations. By composing the transformations of these two groups, we obtain the Poincaré group with 10 parameters

$$x'_i = a_{ik}x_k + a_i \tag{2.21}$$

Poincaré group is relevant in the formulation of Einstein relativity principle. When $c \to \infty$ so that $v/c \ll 1$, Minkowski spacetime reduces to that of Newton and Poincaré group reduces to Galileo group. Let us synthesize the geometric structure of spacetime groups in the following scheme:

$$\text{GALILEO} \Rightarrow \begin{cases} Rotations\ of\ 3-dimensional\ space \\ Boosts \\ 3-dimensional\ space\ translations \\ Time\ translations \end{cases}$$

$$\text{POINCARE}' \Rightarrow \begin{cases} Rotations\ of\ 4-dimensional\ space \\ 4-dimensional\ translations. \end{cases}$$

2.3 The Fantappié "Erlangen Program"

In 1872 Felix Klein (1849–1925) presented the so-called Erlangen program for geometry, centred upon the symmetry transformations groups. From 1952, Fantappié, basing on a similar idea and in perfect consonance with relativity spirit, proposed an Erlangen program for physics, whose essential idea was to individuate physical laws starting from the transformations group which let them invariant. We observe here that there are infinite possible transformations groups which individuate an isotropic and homogeneous space-time. In order to build the next improvements in physics using the group extension method, we can follow the path indicated by the two groups we know to be valid description levels of the physical world: the Galileo group and the Poincarè group. It is useful to remember that the Galileo group is a particular case of the Poincaré one when $c \rightarrow \infty$, i.e. when it is not made use of the field notion and the interactions velocity is considered to be infinite. Staying within a quadri-dimensional spacetime and consequently considering only groups at 10 parameters and continuous transformations, Fantappié showed [8] that the Poincaré group can be considered a limiting case of a broader group F depending with continuity on c and another parameter r. Moreover, this group cannot be furtherly extended provided that only ten parameters are allowed. So we have the sequence:

$$ G_{1+3}^{10} \quad \rightarrow \quad P_{1+3}^{10} \quad \rightarrow \quad F_{1+3}^{10} $$

where G, P and F are respectively the Galileo, Poincaré and Fantappié (that is, de Sitter) group; when $r \rightarrow \infty$, the latter becomes the P group. It is shown that such sequence is univocal. The F group is the one of the five-dimensional rotations of a new spacetime: the maximally symmetric de Sitter universe at constant curvature $1/r^2$. We point out the de Sitter model is obtained without any reference to the gravitational interaction, differently from the General Relativity where the de Sitter universe is one of the possible solutions of the Einstein equations with cosmological constant. From a formal viewpoint we make recourse to five-dimensional rotations because in the de Sitter universe there appears a new constant r, which can be interpreted as the Universe radius. The group extension mechanism individuates an univocal sequence of symmetry groups; for each symmetry group we have a corresponding level of physical world description and a new universal constant, so providing the most general boundary conditions and constraining the form of the possibile physical laws. The de Sitter group fixes the c and r constants and defines a new relativity for the inertial observers in de Sitter universe. In this sense we actually have a version of what is sought for in the Holographic Principle: the possibility to describe laws and boundaries in a compact and unitary way. In 1956 Giuseppe Arcidiacono proposed to study the de Sitter absolute universe by means of the tangent relative spaces where observers localize and describe the physical events by using the Beltrami-Castelnuovo P_4 projective representation. The Projective Special Relativity (PSR) is thus obtained, which collapses in the usual Special Relativity (SR) when $r \rightarrow \infty$ [9].

2.4 The Arcidiacono Transformations[2]

The de Sitter universe is represented by the surface of a five-dimensional hyperboloid H_4 in real time and a five-dimensional sphere S_4 in imaginary time. Arcidiacono considered the flat projective representation of H_4, giving the spacetime P_4 which generalizes Minkowski spacetime. P_4 is defined as the interior of an hyperboloid with two sheets, with center O and semi-axis ir, called the Cayley-Klein *absolute*. The equation of the sheets is, in usual physical coordinates x, y, z, ict:

$$x^2 + y^2 + z^2 - c^2 t^2 + r^2 = 0 \qquad (2.22)$$

The manifold (2.22) meets the time axis in correspondence of "singularities" $t = \pm t_0$ where $t_0 = r/c$ is the time the light takes to travel the Universe radius r. These "singularities" are purely geometrical, not physical; they are placed on the two sheets of the de Sitter horizon of the observer O.

The de Sitter transformations are represented by the projections that transform the absolute in itself. Let us rewrite Eq. (2.22) as ($x_1 = x$, $x_2 = y$, $x_3 = z$, $x_0 = ict$):

$$x_1^2 + x_2^2 + x_3^2 + x_0^2 + r^2 = 0 \qquad (2.23)$$

Introducing the homogeneous coordinates $\bar{x}_a (a = 1, \ldots, 5)$ so defined ($i = 0, 1, 2, 3$):

$$x_i = r \bar{x}_i / \bar{x}_5 \qquad (2.24)$$

it becomes

$$\bar{x}_1^2 + \bar{x}_2^2 + \bar{x}_3^2 + \bar{x}_4^2 + \bar{x}_5^2 = 0 \qquad (2.25)$$

The coordinate $\underline{x}_5$ is determinated according to the Weierstrass condition:

$$\underline{x}^a \underline{x}_a = r^2 \quad a = 0, 1, 2, 3, 5 \qquad (2.26)$$

The inverse relations of (2.24) are:

$$\underline{x}_i = x_i/A; \, \underline{x}_5 = r/A \qquad (2.27)$$

where $A^2 = 1 + x^i x_i / r^2 = 1 + \alpha^2 - \gamma^2$, with $\alpha = (x^k x_k)^{1/2}/r, k = 1, 2, 3$ and $\gamma = ct/r = t/t_0$. The requested transformation between the two O' and O observers consequently has the form:

$$\underline{x}'_a = A_{ab}\underline{x}_b \tag{2.28}$$

where A is an orthogonal matrix. Following the same standard method also used in SR, we get three families of transformations:

(a) The spatial translations of parameter T along (we say) the x axis, given by the $(\underline{x}_5, \underline{x}_1)$ rotation:

$$
\begin{aligned}
\underline{x}'_0 &= \underline{x}_0 \\
\underline{x}'_5 &= -\underline{x}_1 \sin\theta + \underline{x}_5 \cos\theta \\
\underline{x}'_1 &= \underline{x}_1 \cos\theta + \underline{x}_5 \sin\theta
\end{aligned}
\tag{2.29}
$$

Using (2.29) and putting $tg\theta = T/r = \alpha$, we derive the spacetime transformations with T parameter

$$
\begin{cases}
x' = \frac{x-T}{1+\alpha x/r}, & y' = \frac{y\sqrt{1+\alpha^2}}{1+\alpha x/r} \\
z' = \frac{z\sqrt{1+\alpha^2}}{1+\alpha x/r}, & t' = \frac{t\sqrt{1+\alpha^2}}{1+\alpha x/r}
\end{cases}
\tag{2.30}
$$

Equations (2.30) reduce, for $r \to \infty$, to the well known spatial translations of Galileo and Einstein relativities.

(b) The time translations of parameter T_0 given by the $(\underline{x}_5, \underline{x}_0)$ rotation:

$$
\begin{aligned}
\underline{x}'_1 &= \underline{x}_1 \\
\underline{x}'_5 &= -\underline{x}_0 \sin\theta_0 + \underline{x}_5 \cos\theta_0 \\
\underline{x}'_0 &= \underline{x}_0 \cos\theta_0 + \underline{x}_5 \sin\theta_0
\end{aligned}
\tag{2.31}
$$

Putting $tg\theta_0 = iT_0/t_0 = i\gamma$, we obtain ($k = 1,2,3$):

$$
\begin{cases}
x'_k = \frac{x_k\sqrt{1-\gamma^2}}{1-\gamma t/t_0} \\
t' = \frac{t-T_0}{1-\gamma t/t_0}
\end{cases}
\tag{2.32}
$$

Also (2.32), when $r \to \infty$, is reduced to the known cases of Galileo and Einstein relativities.

(c) The boosts of parameter V along (we say) the x axis, given by the $(\underline{x}_1, \underline{x}_0)$ rotation:

$$
\begin{aligned}
\underline{x}'_5 &= \underline{x}_5 \\
\underline{x}'_0 &= -\underline{x}_1 \sin\theta_1 + \underline{x}_0 \cos\theta_1 \\
\underline{x}'_1 &= \underline{x}_1 \cos\theta_1 + \underline{x}_0 \sin\theta_1
\end{aligned}
\tag{2.33}
$$

Putting $tg\theta_1 = iV/c = i\beta$, we reobtain the Lorentz transformations:

$$\begin{cases} x' = \frac{x - Vt}{\sqrt{1-\beta^2}} & y' = y \\ t' = \frac{t - Vx/c^2}{\sqrt{1-\beta^2}} & z' = z \end{cases} \tag{2.34}$$

The general composition of (a), (b) and (c) transformations forms the generic element of de Sitter projective group which, for two variables (x, t) and three parameters (T, T_0, V), can be written as:

$$x' = \frac{Ax + [\beta + \gamma(\alpha - \beta\gamma)]ct + BT}{A(\gamma\beta - \alpha)x/r + (\gamma - \alpha\beta)t/t_0 + B} \tag{2.35}$$

$$t' = \frac{A\beta x/c + [1 + \alpha(\alpha - \beta\gamma)]t + BT_0}{A(\gamma\beta - \alpha)x/r + (\gamma - \alpha\beta)t/t_0 + B} \tag{2.36}$$

where:

$$A^2 = 1 + \alpha^2 - \gamma^2, B^2 = 1 - \beta^2 + (\alpha - \beta\gamma)^2, \alpha = x/r \quad \gamma = t/t_0 \tag{2.37}$$

For $r \to \infty$, we have $A = 1$ and $B = (1 - \beta^2)^{1/2}$ and, from (2.35,2.36), the Poincaré group with three parameters (T, T_0, V). The De Sitter universe with $1/r^2$ constant curvature shows an elliptic geometry in its hyper-spatial global aspect (Gauss-Riemann) and an hyperbolic geometry in its spacetime sections (Lobacevskij). Making the "natural" r unit of this two geometries tend towards infinity we obtain the parabolic geometry of Minkowski flat space.

2.5 Some Consequences

Let us conider a spatial interval χ on H_4 and its projection x on P_4. In addition, let us consider a time interval τ on H_4 and its projection t on P_4; in both cases, an extreme of the interval coincides with the point-event of observation O.

The point-event O is the vertex of a light-cone and the private spacetime P_4 of O is the portion of this light-cone having the absolute as its boundary.

Let us consider two point-events A and B inside P_4. The usual Pythagorean expression in the difference of A and B coordinates not represents a suitable notion of distance AB, because it is not invariant respect to projections. Given a AB straight line intersecting the absolute in R and S, the projective distance is given by the logarithm of the (ABRS) cross ratio:

$$AB = \left(\frac{t_0}{2}\right) \log(ABRS) = \left(\frac{t_0}{2}\right) \log(AR \cdot BS)/(BR \cdot AS) \qquad (2.38)$$

From the (2.38) we obtain:

$$\chi = r\, arctg\left(\frac{x}{r}\right); \quad \tau = \frac{t_0}{2} \log\left(\frac{t_0 + t}{t_0 - t}\right). \qquad (2.39)$$

Thus χ is bounded while x is not and, viceversa, τ is unbounded while t is. As a consequence of this latter result a new law of addition of durations holds:

$$T = \frac{T_1 + T_2}{1 + T_1 T_2 / t_0^2} \qquad (2.40)$$

It is obtained by (2.32) and finds its physical meaning in the identity of the cronological distance of all the observers from the de Sitter horizon; in other words the "de Sitter time" $t_0 = r/c$ is the same for any P_4 observer, exactly as c.

Even the speed composition law is modified according to de Sitter transformations. Let us suppose a certain object moves respect to O with rectilinear uniform motion of speed V. If O in turn moves respect to the inertial observer O′ with rectilinear uniform motion of speed U, O′ sees the same object in rectilinear uniform motion at the speed W given by:

$$W = \frac{(1+\alpha^2)U + (1 - \gamma^2)V + \alpha\gamma c(1 - UV/c^2)}{A(1 + UV/c^2)} \qquad (2.41)$$

For the visible universe of the observer O′ the condition $\alpha = \pm\gamma$ (that is $A = 1$) holds, and the previous equation can be simplified as:

$$W = \frac{U + V \pm \alpha^2 c(1 + U/c)(1 - V/c)}{1 + UV/c^2} \qquad (2.42)$$

For $V = c$ then $W = c$, according to SR, while for $U = c$ we have:

$$W = c \pm 2\alpha^2 c \left(1 - \frac{V}{c}\right)\left(1 + \frac{V}{c}\right) \neq c \qquad (2.43)$$

Eequation (2.43) expresses the possibility of observing hyper-c velocity in PSR. The outcome is less strange than it can seem at first sight, because now the space-time of an observer is defined not only by the c constant but also by r, and the light-cone is at variable aperture. In fact from relations $A^2 = 1 + \alpha^2 - \gamma^2$, $B^2 = 1 - \beta^2 + (\alpha - \beta\gamma)^2$ the following reality condition for B immediately follows:

$$1 - \beta^2 + (\alpha - \beta\gamma)^2 \geq 0 \tag{2.44}$$

which is a quadratic inequality in β:

$$\left(1 - \gamma^2\right)\beta^2 + 2\alpha\gamma\beta - \left(1 + \alpha^2\right) \leq 0 \tag{2.45}$$

or

$$\frac{-\alpha\gamma - A}{1 - \gamma^2} \leq \beta \leq \frac{-\alpha\gamma + A}{1 - \gamma^2} \tag{2.46}$$

In straighter physical terms it means that when we observe a far region of Universe placed at a chronological distance comparable to $t_0 = r/c$, the relative speed of cosmic objects within that region can exceed c value, even if the region belongs to our past light-cone. For $B = 0$ we obtain the angular coefficients of the tangents to the Cayley-Klein absolute exiting from a point P of P_4, representing the two extreme straight lines of the light-cone with vertex P. Differently from Special Relativity, the light-cone opening angle is not constant and depends on the point P according to the formula:

$$tg\theta = 2A/\left(\alpha^2 + \gamma^2\right) \tag{2.47}$$

From the (2.46) derives the variation of the light velocity C with time (c = local value):

$$C = \frac{c}{\sqrt{1 - \gamma^2}} \tag{2.48}$$

from which follows that $C \to \infty$ in the two singularities $t = \pm t_0$.

Exactly as in Special Relativity, the enlargement of the symmetry group involves a deep redefinition of mechanics. In PSR, the m mass of a body varies with velocity and distance according to

$$m = m_0 \frac{A^2}{B} \tag{2.49}$$

From (2.49) it follows that on the lightcone ($A = 0$) the mass is null, while on the absolute ($B = 0$) $m \to \infty$. The local mass of a body at rest varies with t according to:

$$m = m_0(1 - \gamma^2) \tag{2.50}$$

and it vanishes for $\gamma = \pm 1$, at the initial and respectively final instant. So the overall picture for an inertial observer in a De Sitter spacetime is that of a Universe coming into existence in a singularity at $-t_0$ time, expanding and collapsing at t_0 time and where the light speed c is only locally constant. In the initial and final instants this speed is infinite and the mass of a given object is zero. In the projective scenario the space flatness is linked to the observer geometry in a universe at constant curvature. All this is linked to the fact that in PSR the translations and rotations are indivisible. In the singularities there is no "breakdown" of the physical laws because the global spacetime structure is univocally individuated by the group which is independent of the matter-energy distribution. In this case, the singularities in P_4 are—more properly—a horizon of events for the observer (de Sitter horizon).

In order to avoid misunderstandings is necessary take into account that x and t are distances from the observer O, and *the dependence of physical quantities on these distances cannot be verified by that observer by means of local measurements*. Locally, the SR theory is fully valid. Therefore, the de Sitter generalization of well known SR relations is purely formal and physically detectable consequences only exist for the propagation of signals over cosmological distances. This argument is detailed in successive chapters.

References

1. Schutz, B.F.: Geometrical Methods of Mathematical Physics. Cambridge University Press, Cambridge (1990)
2. Wigner, E.: Group Theory and its Applications to the Quantum Mechanics of Atomic Spectra. Academic Press, New York (1959)
3. Arcidiacono, G.: Collectanea Math. **XV**, 259–271 (1963)
4. Gilmore, R.: Lie Groups, Physics, and Geometry: An Introduction for Physicists, Engineers and Chemists. Cambridge University Press, Cambridge (2008). 2008
5. Cohn, P.M.: Lie Groups (Cambridge Tracts in Mathematical Physics). Cambridge Press, Cambridge (1957)
6. Weyl, H.: The Classical Groups. Their Invariants and Representations. Princeton Press, Princeton (1946)
7. Weyl, H.: The Theory of Groups and Quantum Mechanics (translated from German by H. P. Robertson). Dover Publication, New York (1950)
8. Fantappié, L.: Su una nuova teoria di relatività finale. *Rendiconti Accademia dei Lincei* serie **8**, fasc. XVII (1954)
9. Arcidiacono, G.: Collectanea Math. **X**, 65–124 (1958)
10. Arcidiacono, G.: Projective Relativity, Cosmology and Gravitation. Hadronic Press, Nonantum (USA) (1986)
11. Arcidiacono, G.: The theory of hyper-spherical universes. International Center for Comparison and Synthesis, Rome (1987)
12. Arcidiacono, G.: Hadron. J. **16**, 277–285 (1993)
13. Arcidiacono, G.: La teoria degli universi. I-II, Di Renzo, Rome (2000). (in Italian)
14. Licata, I.: El. J. Theor. Phys. **10**, 211–224 (2006)

Chapter 3
Projective Special Relativity

The structural facets of projective relativity are there investigated, in particular the physical meaning of hypersphere and its projection centered on an observer.

3.1 Preliminary Considerations

Let us represent the motion of a material point using a system of three spatial coordinates x and a time coordinate t; the elementary motion will thus be defined by the displacement $(x, t) \rightarrow (x + dx, t + dt)$ or, alternatively, by the quantities (x, t), $v = dx/dt$ and dt.

In (x, t) the vector v undergoes to a change due either to the bodies and fields acting on the material point in (x, t), or merely to the choice of coordinates (e.g. centrifugal effects in a rotating frame). One can therefore write, without losing in generality:

$$dv' = f_1(x', v', \tau)d\tau + f_2 \qquad (3.1)$$

where v', x', τ are defined starting from x, v, t as will be specified below, and $dx' = v'd\tau$. The function f_1 describes the action on v' of the bodies and fields present in (x, t) while f_2 (function of x, t, dx, dt) is a contribution to dv' derived from the choice of coordinates. If the identity $f_2 = 0$ holds, the reference frame where x, t, dx, dt are evaluated is called inertial; in this frame v' is constant with respect to τ when $f_1 = 0$, i.e. when no interaction is present.

If we postulate the existence of at least one inertial frame (1st principle of dynamics), the law (2nd principle of dynamics)

$$dv'/d\tau = f_1(x', v', \tau) \qquad (3.2)$$

must apply within it.

© The Author(s) 2017
I. Licata et al., *De Sitter Projective Relativity*, SpringerBriefs in Physics,
DOI 10.1007/978-3-319-52271-5_3

The vector f_1 is the force per unit mass acting on the material point. If the material point is free, $f_1 = 0$ and therefore v' is constant in τ; consequently, the function $x'(\tau)$ is linear. This result is naturally valid for every system of coordinates (x, t) provided that it is inertial. The passage from one inertial system of coordinates to another thus has the effect of transforming a linear trajectory $x' = x'(\tau)$ into a new linear trajectory. This transformation of coordinates on the space $\{x, t\}$ must therefore induce a *linear* transformation of coordinates on the space $\{x', \tau\}$.

Now, the most general linear transformation of coordinates is the projectivity, so that, given a reference frame in the space $\{x', \tau\}$ associated with an inertial system of coordinates (x, t), the generic projectivity changes this reference into a new reference which is also associated with an inertial system of coordinates.

Two elementary motions relating to the same material point or to different material points can occur in two different locations in the space $\{x', \tau\}$; in this case, a causal connection between them is possible only if a timelike universe line exists that joins such locations, and therefore a proper time interval T' (counted along this line) that separates them. In the case that the two events (relating to different material points) coincide in x' and τ, they can be distinguished by the value of v'; in this latter case, they are only separated by a relative velocity vector of modulus Δ' and there is no impediment to the possibility of a causal link between them.

The crucial observation at this point is that the only invariant of the projectivity consists of the cross-ratio. Therefore, in order to express an invariant law for transformations connecting inertial reference frames (as is the 2nd principle of dynamics) through the parameters T' e Δ', these parameters must be represented by cross-ratios. It is now possible to define the passage from the original coordinates x, t, v to the parameters x', v', τ to be substituted in the 2nd principle.

Let us first of all postulate that the original space $\{x\}$ is real, Euclidean and three-dimensional and that the distance d from the origin is expressed by Pythagoras' theorem:

$$d^2 = x_1^2 + x_2^2 + x_3^2 \tag{3.3}$$

Let us postulate that the original space $\{v\}$ is real, Euclidean and three-dimensional and that the distance Δ from the origin is expressed by Pythagoras' theorem:

$$\Delta^2 = v_1^2 + v_2^2 + v_3^2. \tag{3.4}$$

Let us postulate that the original time t is a real, one-dimensional, Euclidean space with distance from the origin expressed by $T = |t|$.

Let us consider the universe line on which the two events we discussed above are placed and let Q, P be the representative points of these events. As we have to express the distance between these two points in terms of a cross-ratio that is independent of other "real" points, two extreme "ideal" points on the line must exist which we shall indicate as O and S (Fig. 3.1). The cross-ratio will therefore be expressed by the relation:

Fig. 3.1 The cross-ratio of two point-events QP. The points O, S are to infinity

$$(QPOS) = (OQ/PO)(PS/QS).$$

QP is the proper time interval T between the two events, measured in the space $\{x, t\}$ and, if we choose Q as the origin of time, then $T = t$. OQ is the interval between the event Q and the beginning of time t, while SQ is the interval between Q and the end of time t. The cross-ratio then becomes:

$$(QPOS) = (OQ/SQ)\,(PS/PO) = (OQ/SQ)\,[(SQ{-}t)/(OQ\,{+}t)].$$

By rigidly translating the segment PQ on the line OS the distance (i.e., the time interval τ) T' between the two events Q and P must remain unchanged, because we are attempting to define a concept of distance which is independent of position. Therefore T' cannot depend explicitly on Q but only on t. It follows that SQ and OQ must be independent of Q and equal to two constants t_0', t_0''. A second observation is that the value of T' cannot depend on which of the two extremes O, S we choose as the beginning (end) of time t. It follows that the two constants t_0', t_0'' must be equal. From here onwards, let $t_0' = t_0'' = t_0$. Hence:

$$(QPOS) = [(t_0{-}t)/(t_0+t)].$$

For the purpose of having an additive definition of distance, let:

$$T' = k\log(QPOS) = k\log[(t_0 - t)/(t_0+t)].$$

The value of k can be found by taking into consideration the limit of this expression for $t_0 \to \infty$. We have:

$$
\begin{aligned}
T' &= -k\log[(t_0+t)/(t_0-t)] \to \\
&\quad - k\log[(1+t/t_0)(1+t/t_0)] \\
&= -2k\log(1+t/t_0) \approx -2k\log exp(t/t_0) = -2k\,t/t_0.
\end{aligned}
$$

Let $k = -(t_0/2)$; we have $T' = t$ and, placing the origin of proper time in point Q also in the space $\{x', \tau\}$, we have $T' = \tau$, so that in conclusion:

$$\tau = (t_0/2)\log[(t_0+t)/(t_0 - t)]. \tag{3.5}$$

If, on the other hand, the two events coincide on the space $\{x', \tau\}$ but are separated by a relative velocity with modulus Δ', then the line that joins them must be imagined on the one-dimensional space of the relative velocity vector moduli. In this case QP = Δ', while OQ and SQ are the differences between the current velocity of the material point (represented, say, in Q) and the two limit velocities in opposite directions. Again, the distance Δ must not depend explicitly on Q nor on the sign adopted for the velocities, thus OQ = SQ = c, where c is a constant independent of Q. Thus:

$$\Delta = k \log(QPOS) = k \log[(c - \Delta')/(c + \Delta')] = -k \log[(c + \Delta')/(c - \Delta')];$$
$$d\Delta = -2ck \, d\Delta'/(c^2 - \Delta'^2).$$

Since Q is chosen as the origin of the velocity space, $\Delta' = v'$ while $\Delta = v$. In strict conformity with the case of the distances, let $k = -c/2$; we therefore obtain:

$$dv = dv'/\left(1 - v'^2/c^2\right) \tag{3.6}$$

We notice that if the space is isotropic (which is certainly true if it is empty), the maximal velocity c cannot depend on the direction of the relative velocity vector of the elementary motion events Q, P. In other words, the limit speed c must be the same in all spatial directions. The meaning of expressions (3.5), (3.6) will be discussed later.

3.2   The Principle of Relativity

In relation (3.6) we have $v' \in (-c, +c)$ for any inertial reference; it follows that the region of modified spacetime $\{x', \tau\}$ accessible by Q is that which falls within the cone $|v'| = |dx'/d\tau| \leq c$. If we momentarily forget relation (3.5), this result means that the transformations between inertial references must preserve the quadric:

$$c^2\tau^2 - x_1'^2 - x_2'^2 - x_3'^2 = 0. \tag{3.7}$$

As is known, their set constitutes the Poincaré group, which admits the general invariant:

$$c^2\tau^2 - x_1'^2 - x_2'^2 - x_3'^2 = c^2 t^2. \tag{3.8}$$

These arguments must now be extended to include (3.5). The time t that appears in (3.8) is that measured by a local clock in a state of rest (proper time) and therefore coincides with the time t of (3.5). Analysis of (3.5) shows that t is included between the extreme values $-t_0$ and $+t_0$; we must therefore have $t^2 \leq t_0^2$ or, from (3.8):

$$c^2\tau^2 - x_1'^2 - x_2'^2 - x_3'^2 - r^2 \leq 0,$$

where $r = ct_0$. To take into account Eq. (3.5), the passage is therefore required to new group of transformations, more extensive than the Poincaré group, which leaves unchanged the quadric:

$$c^2\tau^2 - x_1'^2 - x_1'^2 - x_1'^2 - r^2 = 0. \tag{3.9}$$

As is known, this is the de Sitter group, admitting the Poincaré group as its limit for $r \to \infty$.

One may wonder about the reason for the transformation of coordinates $(x, v, t) \to (x',v',\tau)$. It is the following: the motion Eq. (3.2) must be valid in every inertial reference system, and must therefore stay valid under transformations that give rise to the passage from one inertial reference system to another. If the "native" Galilean coordinates (x, v, t) were to be adopted, we would notice that the left-hand member of (3.2) actually preserves its form under the relevant group of transformations (Galileo group), but this is not generally true for the right-hand member. For example, electromagnetic interactions do not preserve their shape under Galileo transformations. The passage to the new coordinates (no longer Euclidean-Galilean) x', τ and to the new impulse expression transforms the Galileo group into the Poincaré group, thus allowing the invariance group of the left-hand member of (3.2) to be made homogeneous with the invariance group of the right-hand member. Now both members of this equation preserve their form under the same group of transformations, which is the Poincaré group. This ensures that (3.2) is actually valid in every inertial system.

If only one force field existed in the physical world, it would always be possible in principle to perform a transformation of coordinates of the type mentioned above, so that the mechanics would be made to conform with the covariance group of that field's equations. This would not involve any new physical principle, but only a useful convention. Yet, in the physical world several distinct interaction fields exist and the surprising experimental result is that the transformation of coordinates required to make (3.2) valid in every inertial reference is the same regardless of the fields appearing in the right-hand member. This fact, which is formally expressed by the foregoing reasoning, is a physical principle of the greatest importance: the principle of relativity. The de Sitter group is the covariance group of all physical laws.

The theory of relativity based on the assumption that physical laws are invariant with respect to the Poincaré group is Special Relativity (SR), while that which assumes that the laws are invariant with respect to the de Sitter group is Projective Special Relativity (PSR). SR is the limit case of PSR for $r \to \infty$.

We owe this manner of introducing the relativity principle to Tyapkin [1]. It has the merit of doing away with the mystery of the constancy of c and t_0 for all observers; indeed, there is nothing mysterious about this constancy. A reference frame is defined as inertial when the impulse and the energy of a free material point are conserved within it, so that the inertial reference definition is closely connected to that of impulse and of kinetic energy. The form of the impulse which appears in the expression of the second principle of dynamics depends, in turn, on how an external interaction modifies the motion of the body; it is a description of the body's inertial properties with respect to its interaction with external fields. An inertial system is therefore defined as such by the properties of the interactions, not by space and time properties. Different approximations in the study of interactions between bodies and fields give rise, therefore, to different descriptions of the inertia of bodies and, consequently, to different concepts of an inertial system, i.e. to different theories of relativity. But space and time remain basically Euclidean and Galilean, see (3.3) and (3.4).

Consequently, c and t_0 are properties of interactions, not of space or time; and, indeed, they are connected with the limit velocity of propagation of the influence generated by an interaction and with the maximum duration of this influence. The passage from the Euclidean-Galilean to the PSR or SR chronotope corresponds to the choice of a particular description of inertia and its purpose is to replace Galilean calculation of the coordinates with a new calculation that makes the two members of (3.2) covariant with respect to the same group of transformations. The coordinates redefined in this way constitute a chronotope, which is no longer Galilean-Euclidean, incorporating c and t_0 as its geometric features. The metric element of this chronotope is an invariant of the symmetry group of (3.2) and by adopting its coordinates the parameters c and t_0 become constants that are the same for all observers. It must be noted that while the choice of the calculation of coordinates is totally conventional in itself, the possibility of choosing a calculation that satisfies covariance requirements for all fields appearing in f_1 is an objective property of nature (principle of relativity). From this point of view, Lorentz and Poincaré vision of relativity was therefore correct; they persisted, however, in identifying the fundamental Euclidean-Galilean description with a useless and impossible etheric material "fluid".

The approach adopted here also makes the passage from the "special" theories introduced (SR and PSR) to the corresponding "general" theories (GR = General Relativity and PGR = Projective General Relativity) which include gravitation more comprehensible. It is plausible that, once a given "special" theory (SR or PSR) has been established, the definition of the local class of inertial systems changes from the pointevent P to an infinitely close pointevent P′. This means stating that the value of the free material point impulse changes in passing from P to P′; i.e., the body undergoes "spontaneous" acceleration. If this acceleration is independent of the mass of the body (Galilei principle) then it can be incorporated in a suitable "deformation" of the chronotope, thus obtaining the corresponding general theory (GR or PGR respectively). With this passage a calculation of coordinates is adopted which is dependent upon the position P, in such a way as to incorporate the effects of the acting field (gravitation) in the free motion. Yet the spacetime reference structure is and remains Euclidean-Galilean: true spacetime does not curve and does not undergo any distortion due to the matter present within it.

Some remarks:

(a) One can observe that the existence of limiting surfaces and the principle of inertia are closely interlinked facts. In informal terms, we request a spacetime where the universe lines of free material points are straight. Thus every transformation of spacetime that changes lines into lines leaves the dynamics unchanged. The more general collineation, on the other hand, is a projectivity, so that these transformations are actually projective transformations. Projective transformations can convert "real" points into "ideal" points, and the existence of limiting surfaces such as the quadric described by (3.9) follows from this.

(b) A similar argument can be applied to velocity vector space in SR. The more general transformation of a vector into a vector is again a projectivity. The limiting surface in this space will be physically associated with the existence of a limit propagation velocity. Thus, the appearance of such velocity is in no way magic or surprising.

3.3   Hyperspheric Prespace

Let us consider, in 5-dimensional real Euclidean space, the surface represented by the equation:

$$r^2 = w^2 - c^2 t^2 + x^2 + y^2 + z^2 \tag{3.10}$$

where $r = ct_0$ is a constant. Using the origin as the centre of projection, let us project the points of this surface onto the plane tangent at a generic point Q $(x = y = z = t = 0)$ belonging to it. This plane (Castelnuovo chronotope) contains both a light cone with origin in Q and the two limiting surfaces (past and future with respect to Q) whose equation is given by (3.9), provided that $x_1 = x$, $x_2 = y$, $x_3 = z$, $\tau = t$. By applying the substitution (Wick rotation) $t \rightarrow jt$, (3.10) becomes the equation of the 5-spherical surface with radius r having its centre at the origin. Arcidiacono showed that the de Sitter group is isomorphic to the group of rotations of this surface around its centre. These rotations change, if the tangent plane is kept fixed, or the point Q or the reference system associated with it. In the first case, there is a passage from an observation pointevent Q to a new observation pointevent Q'; in the second case, a rotation or a boost is effected on the reference having its origin in Q. In general, a combination of the two circumstances will occur.

The tangent plane thus constitutes the "private chronotope" where the observer placed in Q coordinates events, while (3.10) defines the "public chronotope" formed by all possible observation events Q. The fifth coordinate is not needed for coordination of events and is therefore not "observable". It is merely a calculation stratagem for linking observations made by different observers. It is real solely in the sense of the intrinsic geometry of Castelnuovo spacetime and not in the extrinsic sense of an actual immersion in the spacetime of a five-dimensional space.

One immediately realises that the theory cannot be generalised to $N > 5$ dimensions. In this case, the hypersphere surface would become $(N - 1)$-dimensional and the plane tangent to it at a given point would also be $(N - 1)$-dimensional.

So if we continued to interpret such a plane as the private spacetime in which an observer associated with the tangent point coordinates events, we would have the appearance of $N - 1 - 4 = N - 5$ physical coordinates in addition to the spacetime ones. Such coordinates, uncompacted, ought to be observable and are not.

We can therefore draw the following conclusion: assuming the first two principles of dynamics and the relativity principle, and assuming an original Galilean-Euclidean ambient space, it follows that an inertial observer coordinates events in a Castelnuovo chronotope. This chronotope can be derived from a 5-dimensional hypersphere through the following succession of commutable operations:

(1) geodesic projection on the 4-dimensional tangent plane passing through the observer (the point of projection is the centre of the hypersphere),
(2) Wick rotation $t \rightarrow jt$.

Reference

1. Tyapkin, A.A.: Expression of the general properties of physical processes in the space-time metric of the special theory of relativity. Sov. Phys. Usp. **15**, 205–229 (1972)

Chapter 4
Point, Fluid and Wave Mechanics

Projective relativity mechanics is here developed in its classical cases and the meaning of the physical quantities in SPR is investigated.

4.1 Introduction

This chapter proposes a modern introduction to point, fluid and wave dynamics, within the context of the theory of Projective Relativity developed by Fantappié (1901–1956) and later by Arcidiacono (1927–1998). We are actually dealing with two distinct theories: the theory of Projective Special Relativity (PSR) and the theory of Projective General Relativity (PGR). The former is a generalization of the ordinary theory of Special Relativity (SR), postulating the invariance of physical laws with respect to the De Sitter group rather than to the Poincaré group, which is a local limit of it [1–7]. The latter is the corresponding generalization of the ordinary theory of general relativity (GR) [8, 9]. The relation between PGR and PSR is the same as that between GR and SR. This chapter will deal exclusively with PSR, which has been restated by various authors under the name of "De Sitter relativity"; it has been discussed in various recent works [10–16].

PSR coincides locally with SR and its only difference from it lies in the predictions relating to the observation of objects that are very distant in space or events that are very distant in time; thus, crucial experiments (or, rather, observations) capable of confuting or verifying PSR can only be carried out in a cosmological context.

In this chapter, the Arcidiacono transformations that generalize ordinary SR Poincaré transformations will not be discussed in detail; for their derivation and properties, the reader is referred to other works [17–21]. After an introduction recalling the kinematics of PSR (Sects. 4.2 and 4.3), the fundamental equations of

© The Author(s) 2017
I. Licata et al., *De Sitter Projective Relativity*, SpringerBriefs in Physics,
DOI 10.1007/978-3-319-52271-5_4

point (Sect. 4.4), perfect incompressible fluid (Sects. 4.5 and 4.7) and wave (Sect. 4.8) dynamics will be introduced. Compared to the original Italian-language works, various deductions have been simplified and some errors have been corrected; also, the physical meaning of equations has been discussed in greater depth. Some comments on the physical meaning of quantities in PSR (Sects. 4.6 and 4.9) have also been added; indeed, this is a topic which can give rise to misunderstandings.

4.2 PSR Metric

In PSR, five projective coordinates, $\underline{x}_0, \underline{x}_1, \underline{x}_2, \underline{x}_3, \underline{x}_5$, are used, which are linked to the physical coordinates x_0, x_1, x_2, x_3, x_5, by the relation:

$$x_i = (\underline{x}_i/\underline{x}_5)r \quad i = 0, 1, 2, 3. \tag{4.1}$$

From here on in this chapter, we shall use the indices $i, j, k, l, m...$ for the values 0, 1, 2, 3 and the indices $a, b, A, B,...$ for the values 0, 1, 2, 3, 5; the Greek indices μ, v will be used when referring only to the spatial coordinates 1, 2, 3. The coordinate x_0 is ict, where t is the chronological distance from an observer, c is the maximum speed of propagation and $i^2 = -1$. The constant r, having the dimensions of one length, is the radius of the de Sitter Universe; the coordinates x_1, x_2, x_3 are the usual spatial coordinates, having their origin in the observer.

Equation (4.1) does not fix the value of $\underline{x}_5$; the Weierstrass condition is assumed:

$$\underline{x}_a \underline{x}^a = r^2. \tag{4.2}$$

Thus, if we pose:

$$A^2 = 1 + \alpha^2 - \gamma^2 = 1 + \alpha_i \alpha^i, \tag{4.3}$$

with $\alpha_i = x_i/r$, $\gamma = ct/r = t/t_0$, it follows from Eq. (4.1) that:

$$\underline{x}_i = x_i/A; \quad \underline{x}_5 = r/A. \tag{4.4}$$

Equations (4.1) and (4.4) allow a coordinate $x_5 = r$ to be introduced; obviously, this is not a physical coordinate in the proper sense of the term, because it is not used by the observer to coordinate events [which occur in the continuum (x_0, x_1, x_2, x_3)]. The introduction of this coordinate facilitates expression of the correlation between data measured by different observers on the PSR chronotope; it must therefore be viewed in the sense of the intrinsic geometry of this chronotope rather than, extrinsically, as a manifestation of its curvature in an "external" five-dimensional space.

The projective metric is:

$$ds^2 = dx_a\, dx^a. \tag{4.5}$$

We observe that $rx_i = x_i x_5$, a relation which, when differentiated, gives:

$$r dx_i = x_i dx_5 + x_5 dx_i. \tag{4.6}$$

By substituting (4.6) into (4.5) we obtain:

$$r^2 ds^2 = (dx_i dx^i)x_5^2 + (r^2 + x_i x^i)dx_5^2 + 2(x_i dx^i)x_5 dx_5,$$

and since $x_5 = r/A$ it follows that:

$$dx_5 = \alpha_i dx^i / A^3 \tag{4.7}$$

$$A^4 ds^2 = A^2 (dx_i dx^i) - (\alpha_i dx^i)^2. \tag{4.8}$$

Equation (4.8) expresses (4.5) in terms of the physical coordinates; it is the metric on the geodetic representation of the de Sitter chronotope (known as the "Castelnuovo chronotope", [22–24]). The fundamental tensor associated to this metric is

$$g_{ik} = (A^2 \delta_{ik} - \alpha_i \alpha_k)/A^4, \tag{4.9}$$

to which corresponds the counter-variant tensor

$$g^{ik} = (A^2 \delta^{ik} - \alpha^i \alpha^k), \tag{4.10}$$

as it can be verified that:

$$g_{is} g^{ks} = \delta_i^k. \tag{4.11}$$

With a tedious but elementary calculation one has:

$$g = \mathrm{Det}(g_{ik}) = A^{-10}. \tag{4.12}$$

The projective D'Alembert operator is thus obtained by using the general formula of mathematical analysis:

$$\Box \varphi = g^{-1/2} \partial_i \left(g^{1/2} g^{ik} \partial_k \varphi \right), \tag{4.13}$$

from which we have:

$$r^2 \Box \varphi = A^2 \left(r^2 \partial_k \partial^k + x^i x^k \partial_i \partial_k + 2x^i \partial_i \right) \varphi. \tag{4.14}$$

For $r \to \infty$, $\Box \to \Box = \partial_i \partial^i$. Wave propagation is described, in PSR, by equations as $\Box \varphi = 0$; this subject will be addressed later from a different viewpoint.

4.3 Kinematics of the Material Point

Equation (4.2) represents the hyperspherical surface of radius r having its centre at the origin, in a 5-dimensional Euclidean space $\{(x_0, x_1, x_2, x_3, x_5)\}$. Let us consider the 4-dimensional space tangent to this hypersphere in a point that coincides with the observer; the hyperspherical surface can be represented on this space by means of a projection from the centre of the sphere (this is known as a "geodetic" projection). Equation (4.8) is thus the Beltrami metric, induced on this space by the projection. This space is called "Castelnuovo chronotope", and it is within it that the observer coordinates events.

Each translation of a material point on the Castelnuovo chronotope is the projection of its motion over the surface (4.2); in other words, each translation on the "physical" chronotope actually is, in the 5-dimensional projective space, a rotation around the origin. Thus, in PSR, translations are a particular class of rotations. This implies that the equation of motion of a material point, rather than assuming the customary Newtonian form $F = dp/dt$, assumes a form which generalizes the equation $L = dM/dt$ valid for rotational motion (F = force, p = impulse, M = angular momentum, L = torque).

From Eq. (4.8), posing $ds = icd\tau$, we have:

$$A^4 d\tau^2 = \left[A^2 (1 - \beta^2) + (\alpha \times \beta - \gamma)^2 \right] dt^2 \tag{4.15}$$

where $\beta = (\beta_0, \beta_1, \beta_2, \beta_3)$, $\beta_\mu = dx_\mu/(cdt)$, $\beta_0 = i$. From the identity

$$(\alpha \times \beta)^2 + (\alpha \wedge \beta)^2 = \alpha^2 \beta^2$$

it thus follows that:

$$A^4 d\tau^2 = \left[(1 - \beta^2) + (\alpha - \beta\gamma)^2 - (\alpha \wedge \beta)^2 \right] dt^2 = \left[B^2 - (\alpha \wedge \beta)^2 \right] dt^2, \tag{4.16}$$

where $B^2 = 1 - \beta^2 + (\alpha - \beta\gamma)^2$. We thus obtain the expression of the proper time interval $d\tau$.[1]

---

[1] In the two-dimensional case (x, t) we have, starting from Eq. (4.15), $A^4 d\tau^2 = B^2 dt^2$.

Thus, the projective velocity:

$$\underline{u}_A = d\underline{x}_A/d\tau \tag{4.17a}$$

and the projective acceleration:

$$\underline{a}_A = d\underline{u}_A/d\tau \tag{4.17b}$$

can be introduced.

From equation $c^2 d\tau^2 = -d\underline{x}_A d\underline{x}^A$ it therefore follows that

$$\underline{u}_A \underline{u}^A = -c^2. \tag{4.18}$$

By deriving Eqs. (4.2), (4.3) with respect to τ we obtain the relations:

$$\underline{x}_A \underline{u}^A = 0; \quad \underline{u}_A \underline{a}^A = 0; \quad \underline{x}_A \underline{a}^A = c^2. \tag{4.19}$$

The projective impulse is defined as:

$$\underline{p}_A = m_0 \underline{u}_A = m_0 d\underline{x}_A/d\tau, \tag{4.20}$$

where m_0 is the local rest mass (i.e. the mass measured by an observer who is at rest with respect to the body *and who occupies the same position as the body*). It follows that:

$$\underline{p}_A \underline{p}^A = -m_0 c^2. \tag{4.21}$$

Let us introduce the physical impulse as:

$$p_i = m_0 u_i = m_0 dx_i/d\tau. \tag{4.22}$$

From equation $x_i = r\underline{x}_i/\underline{x}_5$ it immediately follows that:

$$u_i = dx_i/d\tau = r(\underline{x}_5 \underline{u}_i - \underline{x}_i \underline{u}_5)/\underline{x}_5^2 \tag{4.23}$$

and therefore

$$p_i = r\left(\underline{x}_5 \underline{p}_i - \underline{x}_i \underline{p}_5\right)/\underline{x}_5^2. \tag{4.24}$$

The projective angular momentum is defined as:

$$M_{AB} = \underline{x}_A \underline{p}_B - \underline{x}_B \underline{p}_A. \tag{4.25}$$

By deriving (4.4) with respect to proper time, we have:

$$\begin{cases} A^3 \underline{u}_i = \left(A^2 \delta_{ik} - \frac{x_i x_k}{r^2}\right) u^k \\ A^3 \underline{u}_5 = -\frac{u_i x^i}{r}. \end{cases} \tag{4.26}$$

By inserting (4.26) in (4.25) we have:

$$M_{5i} = r p_i / A^2; \quad M_{ik} = m_{ik}/A^2, \tag{4.27}$$

where $m_{ik} = x_i p_k - x_k p_i$ is the usual physical angular momentum. From (4.25) and (4.19) we obtain:

$$M_{AB} M^{AB} = 2r^2 \underline{p}_A \underline{p}^A. \tag{4.28}$$

Equation (4.28) can be expanded, using (4.27), in the form:

$$-2m_0^2 c^2 A^4 = \frac{m_{ik} m^{ik}}{r^2} + 2p_i p^i \tag{4.29a}$$

or:

$$E = \pm c \sqrt{p^2 + m_0^2 c^2 A^4 + \frac{m_{ik} m^{ik}}{2r^2}}. \tag{4.29b}$$

From (4.27) and (4.19), we also obtain:

$$M_{AB} \underline{u}^A \underline{x}^B = m_0 c^2 r^2, \tag{4.30}$$

while from (4.26) and from $\underline{x}_A \underline{x}^A = r^2$ one obtains:

$$A^3 p_i = p_i - x^k m_{ik}/r^2, \tag{4.31}$$

which is a relation between the impulse and the angular momentum. Finally, the projective moment of inertia tensor is introduced:

$$I_{AB} = m_0 \underline{x}_A \underline{x}_B. \tag{4.32}$$

At small distances from the observer, $\underline{x}_i \approx x_i$ and $\underline{x}_5 \approx r$ whereby, within this limit:

$$I_{ik} = m_0 x_i x_k; \quad I_{i5} = m_0 x_i r; \quad I_{55} = m_0 r^2. \tag{4.33}$$

In other words, the ordinary moment of inertia, the static moment and the mass of the body are combined in I_{AB}.

4.4 Dynamics of the Material Point

The projective torque tensor is defined as:

$$L_{AB} = \underline{x}_A \underline{f}_B - \underline{x}_B \underline{f}_A;$$ (4.34)

in this definition, $\underline{f}_A$ is the projective force vector. Based on what has been said in the previous section, the equation of motion is

$$\frac{dM_{AB}}{d\tau} = L_{AB}.$$ (4.35)

The concept of "free material point" requires some attention. According to (4.35) this type of body is characterized by the condition $L_{AB} = 0$; now:

$$\frac{dM_{5i}}{d\tau} = L_{5i} = \underline{x}_5 \underline{f}_i - \underline{x}_i \underline{f}_5,$$

and for a free point we shall therefore have $L_{5i} = 0$. This condition in no way implies that $\underline{f}_i$ and $\underline{f}_5$ are simultaneously null, and indeed we shall see that they are not.

By virtue of (4.27), the condition $L_{AB} = 0$ becomes:

$$\frac{d}{d\tau}\left(\frac{p_i}{A^2}\right) = 0; \quad \frac{d}{d\tau}\left(\frac{m_{ik}}{A^2}\right) = 0.$$ (4.36)

On the other hand:

$$L_{AB} = \frac{dM_{AB}}{d\tau} = \frac{d}{d\tau}\left(\underline{x}_A \underline{p}_B - \underline{x}_B \underline{p}_A\right) = m_0(\underline{x}_A \underline{a}_B - \underline{x}_B \underline{a}_A),$$

whereby

$$\underline{x}_A \underline{a}_B - \underline{x}_B \underline{a}_A = 0.$$ (4.37)

By multiplying both members of (4.37) by $\underline{u}^A$ and contracting on index A, we obtain the identity $0 = 0$; whereas, by multiplying them by $\underline{x}^B$ we obtain:

$$\underline{a}_A = H^2 \underline{x}_A,$$ (4.38a)

where $H = 1/t_0 = r/c$. From Eq. (4.20) one thus has:

$$d\underline{p}_A/d\tau = m_0 H^2 \underline{x}_A.$$ (4.38b)

Equation (4.38b) splits into the pair of relations:

$$dp_i/d\tau = m_0 H^2 x_i, \quad dp_5/d\tau = m_0 H^2 x_5.$$

By multiplying the first of these by x_5, the second by x_i and subtracting one has:

$$x_5 \frac{dp_i}{d\tau} - x_i \frac{dp_5}{d\tau} = \frac{dM_{5i}}{d\tau} = m_0 \left(x_5 H^2 x_i - x_i H^2 x_5 \right) = 0,$$

and from the first of Eqs. (4.27) one thus obtains the first of Eqs. (4.36) again. Recalling (4.16), it takes the form (V = velocity vector):

$$\frac{d}{dt} \left\{ \frac{m_0 V}{\left[1 - \beta^2 + (\alpha - \beta\gamma)^2 - (\alpha \wedge \beta)^2 \right]^{1/2}} \right\} = 0. \qquad (4.39)$$

The solution of (4.39) is relatively easy in the 2-dimensional case (x, t); it becomes:

$$[1 + \alpha(\alpha - \beta\gamma)] \left(\frac{dV}{dt} \right) = 0. \qquad (4.40)$$

We have two solutions; one is constituted by uniform rectilinear motion V = constant; the other is expressed by $\beta = (1 + \alpha^2)/(\alpha\gamma)$, which can easily be rewritten as:

$$\frac{dx}{dt} = \left(\frac{r}{t} \right) \left(\frac{x}{r} + \frac{r}{x} \right). \qquad (4.41)$$

Equation (4.41) is a differential equation with separable variables whose solution is:

$$x^2 - k^2 t^2 + r^2 = 0, \qquad (4.42)$$

where k is an arbitrary constant. One immediately sees that in the observer's present $(t = 0)$ one has $x = \pm ir$, an imaginary result. To avoid this singularity of kinematics, one must impose that these bodies not be simultaneous to any observer, but this is tantamount to saying that they are not physical. In other words, the solutions of (4.41) are not physically admissible and must be ruled out; with this exclusion, the only possible free motion remaining (in the two-dimensional case) is uniform rectilinear motion.

At this point, a digression is necessary. Let us consider (4.29b) again, which we rewrite in the form

$$p^2 - \frac{E^2}{c^2} + \frac{m_{ik}m^{ik}}{r^2} = -m_0^2 c^2 A^4. \tag{4.43}$$

For a material point at rest, $p = 0$ and $m_{ik} = 0$, so that:

$$E = m_0 c^2 A^2 = m_0 c^2 \left(1 + \frac{x^2}{r^2} - \frac{x_0^2}{r^2} \right) = m_0 c^2 \left(\frac{r^2}{r^2} + \frac{x_\mu x^\mu}{r^2} - \frac{x_0^2}{r^2} \right)$$

$$= m_0 c^2 \left(\frac{x_5 x^5}{r^2} + \frac{x_\mu x^\mu}{r^2} - \frac{x_0^2}{r^2} \right) = m_0 c^2 \frac{x_A x^A}{r^2} = m_0 H^2 x_A x^A,$$

because $x_5 = r$. This, therefore, is the expression of rest energy in PSR. As regards *local* rest energy, it is expressed by:

$$m_0 c^2 = \frac{E}{A^2} = m_0 H^2 \frac{x_A x^A}{A^2} = m_0 H^2 \underline{x_A} \underline{x}^A.$$

Let us therefore assume the following expression for the energy tensor of the free material point:

$$T_{AB} = m_0 \left(\underline{u_A} \underline{u_B} - H^2 \underline{x_A} \underline{x_B} \right). \tag{4.44}$$

In this expression, the term $m_0 H^2 \underline{x_A} \underline{x_B}$, whose spur is equal to local rest energy, is subtracted from the term $m_0 \underline{u_A} \underline{u_B}$ which comes from the direct generalization of the similar SR expression. The term $m_0 H^2 \underline{x_A} \underline{x_B}$ is null in the limit $r \to \infty$, in which SR is re-obtained.

To verify the validity of (4.44), let us define the projective force as[2]:

$$\underline{f_A} = \partial^B T_{AB}. \tag{4.45}$$

For a free material point we therefore have, considering that $\partial^A \underline{x_A} = 5$:

$$\underline{f_A} = -5 m_0 H^2 \underline{x_A} \tag{4.46}$$

i.e.

$$\underline{f_i} - \frac{x_i}{r} \underline{f_5} = -5 m_0 H^2 \left(\underline{x_i} - \frac{x_i}{r} \underline{x_5} \right) = 0, \tag{4.47}$$

---

[2]To avoid unduly complex notation from here on until Sect. 4.7 inclusive, we shall use the symbol ∂_A to indicate the partial derivative with respect to the variable $\underline{x_A}$, rather than with respect to the variable x_A. In Sect. 4.8 the definition of projective derivation will be made explicit, and the related notation $\underline{\partial_A}$ will be introduced.

because $x_i = r\underline{x}_i/\underline{x}_5$. Now, the relation $M_{5i} = dL_{5i}/d\tau$ can be rewritten as:

$$\underline{x}_5\underline{f}_i - \underline{x}_i\underline{f}_5 = \frac{d}{d\tau}\left(\underline{x}_5\underline{p}_i - \underline{x}_i\underline{p}_5\right)$$

and, dividing it by $\underline{x}_5$:

$$\underline{f}_i - \frac{x_i}{r}\underline{f}_5 = \frac{1}{\underline{x}_5}\left(\dot{\underline{x}}_5\underline{p}_i + \underline{x}_5\dot{\underline{p}}_i - \dot{\underline{x}}_i\underline{p}_5 - \underline{x}_i\dot{\underline{p}}_5\right)$$

$$= \dot{\underline{p}}_i - \frac{x_i}{r}\dot{\underline{p}}_5 + \frac{1}{\underline{x}_5}\left(\dot{\underline{x}}_5\underline{p}_i - \dot{\underline{x}}_i\underline{p}_5\right).$$

On the other hand:

$$\dot{\underline{x}}_5\underline{p}_i - \dot{\underline{x}}_i\underline{p}_5 = \left(\frac{\underline{p}_5}{m_0}\right)\underline{p}_i - \dot{\underline{x}}_i\underline{p}_5$$

$$= \underline{p}_5\left(\frac{\underline{p}_i}{m_0}\right) - \dot{\underline{x}}_i\underline{p}_5 = \underline{p}_5\dot{\underline{x}}_i - \dot{\underline{x}}_i\underline{p}_5 = 0,$$

so that:

$$\underline{f}_i - \frac{x_i}{r}\underline{f}_5 = \dot{\underline{p}}_i - \frac{x_i}{r}\dot{\underline{p}}_5.$$

Thus, from (4.47) we have, for a free material point:

$$\underline{\dot{p}}_i - \frac{x_i}{r}\underline{\dot{p}}_5 = 0. \tag{4.48}$$

As a matter of fact, this equation is certainly valid because it can be derived from (4.38b), if we recall that $x_i = r\underline{x}_i/\underline{x}_5$. We can therefore conclude that (4.44) and (4.45) are compatible with the dynamics of the free material point.

It is appropriate to point out that, also in the case of a free material point, we have $f_A \neq 0$, as is clearly evidenced by (4.46). In PSR, it is the torque that is null in the free case, not the force; indeed, the spacetime translations are, in turn, rotations and therefore only rotations exist in reality. In free motion, the time variation of p_5 cancels that of p_i, as is evidenced in (4.48); in the two-dimensional case this implies uniform rectilinear motion.

We can obtain the same result by considering (4.44). The term $-m_0H^2x_Ax_B$ depends on the coordinates: its divergence is therefore a force which, by acting on the point, determines its free motion. This force is precisely the left-hand of (4.46).

One should stress that the conventional treatment of the de Sitter chronotope [25] does not make use of projective coordinates, and therefore p_5 does not exist in that context. Furthermore, the motion equation is assumed to have the form $F = dp/dt$, rather than $L = dM/dt$. In the case of a free material point, this approach leads us to identify the force with expression (4.38b) which, for remote events that can be observed through their light and therefore placed on the observer's lightcone ($\alpha^2 = \gamma^2 \to A^2 = 1$ and $\underline{x}_A = x_A$), becomes $f_\mu = m_0 H^2 x_\mu$. In the conventional treatment, one has $H^2 = \lambda/3$, where λ is the cosmological constant; thus, f_μ is nothing other than the "cosmological term". In other words, the disappearance of the "balancing" term $\underline{p}_5$ leads to a free motion which is no longer uniform but accelerated, and the force that must be introduced as the cause of this acceleration is the cosmological term.

It is possible to make free motion uniform again by suitably re-graduating clocks; this strategy leads to Milne's double time scale [26].

Having spoken of how the "cosmological term", non-existent in PSR, emerges in conventional theory, we ought now to speak of another important aspect of the de Sitter chronotope, to see how it is described in PSR: cosmic expansion.

In PSR, cosmic expansion derives from the transformations of coordinates which change one inertial system into another; these are Fantappié-Arcidiacono transformations, generalizations of Lorentz-Poincaré transformations. It is therefore a *kinematic* and not a dynamic fact; this particular must be borne in mind.

The transformations relevant here are the time translations of parameter T_0; under one of these [17–21], the velocity V of a body located in the event point (x, t) of the unprimed reference frame becomes V', where:

$$V'\sqrt{1 - \gamma^2} = V(1 + \gamma t/t_0) - \gamma x/t_0 \qquad (4.49)$$

and $\gamma = T_0/t_0$. If, in the unprimed reference frame, the body moves with uniform motion according to the law $x = Vt + x_0$ and $\gamma^2 \neq 1$, then

$$V' = \frac{V - \gamma\frac{x_0}{t_0}}{\sqrt{1 - \gamma^2}}, \qquad (4.50)$$

which is a constant. Therefore, even in the primed reference frame the motion will be uniform rectilinear and its velocity will be V'. This is not a property peculiar to time translations but a common feature of all the transformations of the de Sitter-Fantappié-Arcidiacono group: they convert uniform rectilinear motions into uniform rectilinear motions. On the other hand this is nothing but a consequence of the covariance, with respect to that group, of (4.36) and its solutions.

From (4.50) it can be seen that for $\gamma \to \pm 1$, $V' \to \infty$ unless $V = \pm x_0/t_0$, a quantity which can assume a multiplicity of values, as the constant x_0 is arbitrary; in this case, (4.50) gives $V' = 0$. The first member of (4.49) thus tends to zero for $\gamma^2 \to 1$, and we obtain:

$$V = \frac{\gamma x/t_0}{1 + \gamma t/t_0}; \qquad (4.51)$$

with $\gamma = \pm 1$ according to the sign of T_0. Equation (4.51) can be verified by directly substituting $x = Vt + x_0$ and $V = \gamma x_0/t_0$; the identity $V = V$ is obtained for every value of γ, thus also for $\gamma^2 \to 1$. In the $\gamma = +1$ case (past lightcone) we have $V = x_0/t_0$, where x_0 is the position of the body on the observer's simultaneity plane, and:

$$V = H(t)x \tag{4.52}$$

where $H(t) = H/(1 + t/t_0)$, $-t_0 \leq t \leq 0$, $H = 1/t_0$. Equation (4.52) expresses the existence of a velocity field escaping from the observer, whose modulus increases with the distance from the latter; it is therefore a law of cosmic expansion. In the future lightcone ($\gamma = -1$), on the other hand, there is a cosmic contraction which is entirely symmetrical to this expansion, though it is not observable as it is not possible to receive signals from the future.

The not trivial fact is the compatibility between uniform free rectilinear motion and cosmic expansion. The field of velocity (4.52) has been derived from the request for non-divergence of the transformed velocity V'; it plays the role held by the "substratum" in Milne's kinematic relativity [27]. The dynamic Eq. (4.35) determines the local deviations from the "substratum" caused by the action of the forces. All this is unknown in ordinary Special Relativity.

The result obtained can be expressed by saying that the primed reference frame, or the system of bodies at rest with respect to it (for which $V' = 0$) exists if these bodies, in the unprimed reference frame, have velocities distributed in accordance with (4.52); i.e. if a cosmic expansion exists in this second reference frame. However, given that the choice of the primed reference frame is arbitrary, this result is equivalent to stating the existence of a class of observers who observe a cosmic expansion as described by (4.52); this class constitutes the "substratum". It is remarkable that the substratum should appears for merely kinematic (group) reasons, without any physical requirements such as the introduction of an aether might be.

4.5 Dynamics of Perfect Incompressible Fluids

In SR the expression of the energy tensor of the perfect incompressible fluid is:

$$T_{ik} = (\mu + p/c^2)u_i u_k + p\delta_{ik}, \tag{4.53}$$

where μ and p are the density and the pressure of the fluid, respectively, and u_i is its quadrivelocity. The PSR generalization of (4.53) is obvious: one must substitute, in the limit $p \to 0$, the disgregated matter tensor $\mu\, u_i\, u_k$ with $\mu(\underline{u}_A\underline{u}_B - H^2\underline{x}_A\underline{x}_B)$. One thus obtains:

$$T_{AB} = (\mu + p/c^2)\left[\underline{u}_A\underline{u}_B - H^2\underline{x}_A\underline{x}_B\right] + p\delta_{AB}. \tag{4.54}$$

Let:

$$f^2 = \mu + p/c^2, \tag{4.55}$$

and recalling that for a perfect incompressible fluid the equation of state[3]:

$$p = \mu c^2 \tag{4.56}$$

applies, (4.54) becomes:

$$T_{AB} = f^2 \underline{u}_A \underline{u}_B + f^2 c^2 \left[(1/2)\delta_{AB} - (1/r^2)x_A x_B \right]. \tag{4.57}$$

Thus, assuming that:

$$c_A = f \underline{u}_A, \tag{4.58}$$

the energy tensor becomes:

$$T_{AB} = c_A c_B - c^S c_S \left[(1/2)\delta_{AB} - (1/r^2)\underline{x}_A \underline{x}_B \right]. \tag{4.59}$$

It can be postulated [28, 29] that this expression also remains valid in the more general case:

$$c_A = f \underline{u}_A + Q_A; \quad Q_A \underline{x}^A = 0; \quad Q_A \underline{u}^A = 0. \tag{4.60}$$

The relations obtained can be written in another form by introducing the generalized Eckart tensor:

$$\eta_{AB} = \delta_{AB} + (1/c^2)\underline{u}_A \underline{u}_B - (1/r^2)\underline{x}_A \underline{x}_B. \tag{4.61}$$

This tensor is symmetric, and in the proper reference all its components are locally null except the spatial ones $\eta_{\alpha\beta} = \delta_{\alpha\beta}$. It satisfies the conditions:

$$\eta_{AB}\underline{x}^A = 0; \quad \eta_{AB}\underline{u}^A = 0; \quad \sum_B \eta_{AB}\eta_{BC} = \eta_{AC}. \tag{4.62}$$

Equation (4.54) becomes:

$$T_{AB} = \mu \left(\underline{u}_A \underline{u}_B - H^2 \underline{x}_A \underline{x}_B \right) + p\eta_{AB}, \tag{4.63}$$

---

[3]We recall that μ breaks down into a pure mass term μ_0 and into a term dependent upon the specific internal energy ε of the fluid, in accordance with the relation $\mu = \mu_0(1 + \varepsilon/c^2)$. The fluid is incompressible in the sense that in isothermal conditions μ_0 is a constant; whereas p obviously depends on the coordinates through ε. In these circumstances, the spacetime part of the fluid field c_A is, in Einstein's $r \to \infty$ limit and in the absence of external forces, solenoidal [20, vol. II].

while (4.59) becomes:

$$T_{AB} = c_A c_B - c^S c_S \left[\eta_{AB} - (1/2)\delta_{AB} - (1/c^2) u_A u_B \right],\qquad(4.64)$$

an expression which keeps its form when the Einstein's limit $r \to \infty$ is performed.
From (4.62) and (4.63) one has:

$$T_{AB}\underline{x}^A = -\mu c^2 \underline{x}_B; \quad T_{AB}\underline{u}^A = -\mu c^2 \underline{u}_B;\qquad(4.65)$$

in other words, $\underline{x}_A$ and $\underline{u}_A$ are eigenvectors of T_{AB} with eigenvalue $-\mu c^2$.
Furthermore:

$$T_{AB}\underline{x}^A \underline{x}^B = T_{AB}\underline{u}^A \underline{u}^B = -\mu c^2 r^2.\qquad(4.66)$$

The generalized Euler equations are obtained by equating to zero the divergence
of (4.54); posing $f^2 = \mu + p/c^2 = m$ one has:

$$m\underline{a}_B + \underline{u}_B \partial_A \left(m\underline{u}^A \right) + \partial_B p - H^2 m \underline{x}_A \partial^A \underline{x}_B - H^2 \underline{x}_B \partial^A (m\underline{x}_A) = 0.$$

By multiplying this expression by $\underline{u}^B$ and $\underline{x}^B$, respectively, two continuity
equations are obtained:

$$c^2 \partial_A \left(m\underline{u}^A \right) = dp/d\tau\qquad(4.67\text{a})$$

$$c^2 \partial_A \left(m\underline{x}^A \right) = dp/d\rho\qquad(4.67\text{b})$$

where τ is the curvilinear coordinate along the stream line, and ρ is the spatial
distance from the stream line. The expressions of the radial derivative $(d/d\rho = \underline{x}_A \partial^A)$
and of the derivative along the stream lime $(d/d\tau = \underline{u}_A \partial^A)$ have been taken into
account. By substituting (4.67) into the principal equation, the generalized Euler
equation is obtained:

$$m\underline{a}_B + (\underline{u}_B/c^2)dp/d\tau - (\underline{x}_B/r^2)dp/d\rho + \partial_B p = H^2 m\underline{x}_B.\qquad(4.68)$$

All the discussion conducted up to this point is valid for perfect fluids. In the
case of viscous fluids, the term $-\nu V_{AB}$, where ν is the viscosity coefficient and V_{AB}
is the viscosity tensor obtained by directly generalizing the SR one [20], must be
added to the second member of (4.64). One has:

$$2V_{AB} = \eta_{AR}\eta_{BS}\left(\partial^R c^S + \partial^S c^R \right).\qquad(4.69)$$

4.6 Digression on the Concept of Temperature in PSR

Before explaining the fundamental equations of fluid with heat exchange in PSR, it is necessary to stop and discuss the concept of temperature in theories of relativity based on a global symmetry group. It is necessary to eliminate any ambiguity on the physical meaning of temperature as a quantity which will appear in those equations. The general problem of the meaning of physical quantities in PSR will be examined in Sect. 4.9.

Let us place ourselves in the context of ordinary SR, and let T_0 be the temperature of a gas measured by an observer at rest with respect to it; what is the temperature T of this same gas measured by a second observer in uniform rectilinear motion at velocity V with respect to the former? As is well known [30, 31] there are, in SR, three distinct definitions of temperature which correspond to the three distinct laws of transformation:

$$T = T_0\gamma; \quad T = T_0\gamma^{-1}; \quad T = T_0, \tag{4.70}$$

where

$$\beta = V/c; \quad \gamma = 1/\sqrt{1 - \beta^2}.$$

The extension of these laws to the PSR domain is simple and obvious. Firstly, T_0 is the temperature measured by an observer who not only is at rest with respect to the gas, but is also located in the same spacetime region occupied by it. The contraction parameter $\gamma = dt/d\tau$ is generalized by the corresponding PSR quantity:

$$\Gamma = \frac{1 + \alpha^2 - \gamma^2}{\sqrt{1 - \beta^2 + (\alpha - \beta\gamma)^2 - (\alpha \wedge \beta)^2}}, \tag{4.71}$$

where $\beta = V/c$, $\alpha = d/r$, $\gamma = t/t_0$. Here d and t are the parameters of the spacetime translation which transports the first observer into the second.

The temperature T measured by the second observer is therefore, in accordance with the three distinct definitions:

$$T = T_0\Gamma; \quad T = T_0\Gamma^{-1}; \quad T = T_0. \tag{4.72}$$

It must be borne in mind [32] that the first "observer" is actually a thermometer, which must be in thermal equilibrium with the gas. Thus, it must be *at rest* with respect to the gas and *immersed* in it; the reading of this thermometer is therefore T_0. Even if the thermometer is read by an observer in motion with respect to it, or placed at cosmological distances from it, the result of the reading will always be T_0. Thus, if T is understood as a "thermometer reading", one must necessarily have

$T = T_0$. This supports the third definition (local proper temperature) and we shall use this one from now on.

4.7   Dynamics of Perfect Incompressible Fluids with Heat Exchange

Arcidiacono studied, both in SR and in PSR, the case of a perfect incompressible fluid (described only by a single index f) subject to heat exchanges. He postulated the relation [28, 29]:

$$c_A = f u_A + Q_A \tag{4.73}$$

with $Q_A \neq 0$, so that the hydrodynamic field c_A is no longer parallel to the fluid stream u_A. Precisely:

$$Q_A = q_A / fc^2, \tag{4.74}$$

where q_A is the so called "thermal vector"; it satisfies the two conditions:

$$q_A \underline{x}^A = 0, \quad q_A \underline{u}^A = 0. \tag{4.75}$$

The thermal vector is linked to the absolute temperature T, defined in accordance with the previous section, by the generalized Fourier equation:

$$q_A = -\chi \eta_{AB} \partial^B T = -\chi \left[\partial_A T + \frac{u_A}{c^2} \frac{dT}{d\tau} - \frac{x_A}{r^2} \frac{dT}{d\rho} \right]. \tag{4.76}$$

In this equation, χ is the thermal conductivity coefficient, which we shall assume to be constant. By substituting (4.73) into (4.59), the energy tensor is obtained:

$$
\begin{aligned}
T_{AB} = {}& f^2 \underline{u}_A \underline{u}_B + \frac{1}{c^2} \left(\underline{u}_A q_B + \underline{u}_B q_A \right) \\
& + \frac{q_A q_B}{f^2 c^4} + \left(f^2 c^2 - \frac{q^2}{f^2 c^4} \right) \left(\frac{\delta_{AB}}{2} - \frac{x_A x_B}{r^2} \right),
\end{aligned}
\tag{4.77}
$$

in which it has been posed $q^2 = q_A \, q^A$. We note that in the non-thermal case ($q_A = 0$) one has $c_A c^A = (f \underline{u}_A)(f \underline{u}^A) = -f^2 c^2 = -(\mu c^2 + p) = -2p$. Assuming the validity of the normalization $c_A c^A = -2p$ for $q_A \neq 0$, as well, one has:

$$f^2 c^2 - \frac{q^2}{f^2 c^4} = 2p. \tag{4.78}$$

Recalling that $f^2 = m = \mu + p/c^2$ one can eliminate f^2 from Eq. (4.78), obtaining:

$$p^2 = \mu^2 c^4 - q^2/c^2, \tag{4.79}$$

a relation similar to that which applies in the relativistic hydrodynamics of SR.

Introducing the tensor Q_{AB} by means of the expression:

$$c^2 Q_{AB} = \underline{u}_A q_B + \underline{u}_B q_A + q_A q_B/mc^2, \tag{4.80}$$

Equation (4.77) becomes:

$$T_{AB} = m\underline{u}_A \underline{u}_B - (2p/r^2)\underline{x}_A \underline{x}_B + p\delta_{AB} + Q_{AB}. \tag{4.81}$$

By equating to zero the divergence of Eq. (4.81) one obtains:

$$m\underline{a}_B + \underline{u}_B \partial^A(m\underline{u}_A) + \partial_B p - \underline{x}_B \partial^A\left[(2p/r^2)\underline{x}_A\right] + \\ - (2p/r^2)\underline{x}_B + \partial^A Q_{AB} = 0. \tag{4.82}$$

By multiplying this expression by $\underline{u}^B$, and bearing in mind that:

$$\underline{u}_B \underline{a}^B = 0; \quad \underline{u}_B \underline{u}^B = -c^2; \quad \underline{u}_B \partial^B p = dp/d\tau; \quad \underline{x}_B \underline{u}^B = 0;$$

the continuity equation is obtained:

$$c^2 \partial^A(m\underline{u}_A) = dp/d\tau + \underline{u}^B \partial^A Q_{AB}. \tag{4.83}$$

Whereas by multiplying (4.82) by $\underline{x}^B$ and recalling that:

$$\underline{x}^B \underline{a}_B = c^2; \quad \underline{x}^B \underline{u}_B = 0; \quad \underline{x}^B \partial_B p = dp/d\rho; \\ \underline{x}_B \underline{x}^B = r^2; \quad \underline{x}_A \partial_A \underline{x}^B = \underline{x}_B; \quad 2\underline{x}^B \partial^A \underline{x}_B = \partial^A r^2 = 0,$$

one obtains:

$$r^2 \partial^A (2p\underline{x}_A/r^2) = mc^2 + dp/d\rho + \underline{x}^B \partial^A Q_{AB} - 2p. \tag{4.84}$$

By substituting (4.83) and (4.84) into (4.82) one obtains:

$$m\underline{a}_B + (\underline{u}_B/c^2)dp/d\tau - (\underline{x}_B/r^2)dp/d\rho + \partial_B p - \partial^A Q_{AB} = H^2 m\underline{x}_B. \tag{4.85}$$

This is the Euler equation proposed by Arcidiacono for perfect incompressible fluids with thermal exchange. When the thermal vector vanishes, this equation is reduced to (4.68).

4.8 D'Alembert Equation

In Sect. 4.2 the D'Alembert projective operator, which rules free wave propagation, was introduced starting directly from the metric. In this section, we propose a different and instructive construction, starting from the projective derivatives [20, vol. II].

By differentiating the equation:

$$x_a = (\underline{x}_a/\underline{x}_5)r \tag{4.86}$$

one obtains:

$$\frac{d\underline{x}_a}{r} = -\frac{\underline{x}_a}{\underline{x}_5^2}d\underline{x}_5 + \frac{d\underline{x}_a}{\underline{x}_5},$$

i.e.:

$$\frac{\partial \underline{x}_a}{\partial \underline{x}_5} = -r\frac{\underline{x}_a}{\underline{x}_5^2}; \quad \frac{\partial \underline{x}_s}{\partial \underline{x}_a} = \frac{r}{\underline{x}_5}\delta_{sa}. \tag{4.87}$$

Let us define the projective derivation with respect to index a as:

$$\underline{\partial}_a = (\partial/\partial\underline{x}_a) = \sum_s \left(\frac{\partial}{\partial x_s}\right)\left(\frac{\partial x_s}{\partial \underline{x}_a}\right).$$

For $a \neq 5$ one has:

$$\underline{\partial}_a = \sum_s \left(\frac{\partial}{\partial x_s}\right)\delta_{sa}\frac{r}{\underline{x}_5} = \left(\frac{\partial}{\partial x_a}\right)\frac{r}{\underline{x}_5} = A\partial_a, \tag{4.88}$$

where $A = r/\underline{x}_5$. For $a = 5$ one has:

$$\begin{aligned}
\underline{\partial}_5 &= \sum_s \left(\frac{\partial}{\partial x_s}\right)\left(\frac{\partial x_s}{\partial \underline{x}_5}\right) = \sum_s (\partial_s)\left(-r\frac{\underline{x}_s}{\underline{x}_5^2}\right) \\
&- A\sum_s (\partial_s)\left(\frac{\underline{x}_s}{\underline{x}_5}\right) = -\frac{A}{r}\sum_s (\partial_s)\left(r\frac{\underline{x}_s}{\underline{x}_5}\right) \\
&= -\frac{A}{r}x_s\partial^s.
\end{aligned} \tag{4.89}$$

In practice, the ordinary partial derivative with respect to the index $s = 5$ is the derivative with respect to the constant $x_5 = r$, and therefore it does not exist. The relations (4.88) and (4.89) express the projective derivatives as a function of the ordinary ones. For $s = 0, 1, 2, 3$ one has:

$$\partial_s\partial^s = (A\partial_s)(A\partial^s) = A(\partial_s A)(\partial^s) + A^2\partial_s\partial^s,$$

and since $\partial_s A = x_s/(Ar^2)$,

$$\partial_s\partial^s = \frac{x_s}{r^2}\partial^s + A^2\partial_s\partial^s.$$

Instead:

$$\partial_5^2 = \left(-\frac{A}{r}x_l\partial^l\right)\left(-\frac{A}{r}x^m\partial_m\right)$$

$$= \frac{A^2}{r^2}x_l\partial^l + \frac{A^2}{r^2}x_l x_m\partial^l\partial^m + \frac{x_l x^m x^l}{r^4}\partial_m,$$

an expression in which the indices l, m run along 0, 1, 2, 3. Since $x_l\,x^l = r^2(A^2 - 1)$, one has:

$$\partial_5^2 = \frac{A^2}{r^2}x_l\partial^l + \frac{A^2}{r^2}x_l x_m\partial^l\partial^m + \frac{(A^2-1)}{r^2}x^m\partial_m.$$

At this point, the projective Dalembertian can be introduced:

$$\Box\varphi = \partial_a\partial^a\varphi = (\partial_s\partial^s\varphi + \partial_5\partial^5\varphi); \quad s = 0, 1, 2, 3.$$

One immediately obtains:

$$\Box\varphi = (A^2/r^2)(r^2\partial_s\partial^s + x_l x_m\partial^l\partial^m + 2x_s\partial^s)\varphi. \qquad (4.90)$$

Equation (4.90) links the projective Dalembertian to the ordinary one $\partial_s\partial^s$. The D'Alembert wave equation thus takes the De Sitter-covariant form:

$$\Box\varphi = 0, \qquad (4.91)$$

and in this form has been extensively studied by Arcidiacono and Capelas de Oliveira [33, 34].

It is to be noted that the components of the wave number vector $\underline{k}_a$ must be appropriately redefined in PSR. The plane wave $\exp(i\underline{k}_a\underline{x}^a)$ is a solution of (4.91) only if $\underline{k}_a\underline{k}^a = 0$. If $\underline{k}_0 = i\omega/c$ is defined as in SR, one must have that $\underline{k}_5 = \theta/r$, if one wants this component to disappear in the limit $r \to \infty$. The condition will then be satisfied if one lets:

$$\underline{k}_\alpha = n_\alpha\left[(\omega/c)^2 - (\theta/r)^2\right], \quad \alpha = 1, 2, 3, \qquad (4.92)$$

with $n_\alpha n^\alpha = 1$. The phase thus becomes $\underline{k}_a\underline{x}^a \to \mathbf{k}\cdot\mathbf{x} - \omega t + \theta$ for $r \to \infty$.

The static case, in which φ not depends on time (i.e. on x_0), is very interesting. In this case, (4.91) becomes the generalized Poisson equation:

$$\Delta \varphi = \left[\partial_\alpha \partial^\alpha + \left(x_\beta x_\gamma / r^2 \right) \partial^\beta \partial^\gamma + \left(2 x_\alpha / r^2 \right) \partial^\alpha \right] \varphi = 0, \qquad (4.93)$$

where the Greek indices run along the ordinary spatial coordinates. In the case of a central field $\varphi = \varphi(\rho)$, $\rho = (x_\alpha x^\alpha)^{1/2}$, this equation admits of the solution [20, 35]:

$$\varphi = -kY(\rho)/\rho, \qquad (4.94a)$$

with

$$Y(\rho) = \left(1 + \rho^2 / r^2 \right)^{1/2} \{ \cos[\mathrm{arctg}(\rho/r)] + \sin[\mathrm{arctg}(\rho/r)] \}. \qquad (4.95b)$$

Note that for $r \to \infty$, $\varphi \to -k/\rho$, and this allows the constant k to be physically identified. For example, in the case of the gravitational field it is clearly the mass of the attracting body, multiplied by the Newton gravitational constant.

From (4.92), with the subsidiary conditions $k_4 = i\omega/c$, $k_0 = \theta/r$ the phase function assumes the form:

$$\varsigma = k_A x^A; \quad k_A k^A = 0. \qquad (4.96)$$

From the transformations of the de Sitter group we obtain the new law of the Doppler effect which takes the form [for $\theta = 0$, $k_1 = (\omega/c)\cos \varsigma$, $k_2 = k_3 = 0$]:

$$\begin{cases} \omega' = \omega[1 + A\beta \cos \varsigma + \alpha(\alpha - \beta\gamma)]/(AB) \\ \theta' = \omega t_0[(\beta\gamma - \alpha)A \cos \varsigma + (\gamma - \alpha\beta)]/(AB) \end{cases} \qquad (4.97)$$

In the case of a radial recession motion this equation gives:

$$(1+z)^{-1} = \sqrt{\frac{1-\beta}{1+\beta} + \alpha^2} \qquad (4.98)$$

where z is the red-shift, that is $1 + z = \omega'/\omega$. Since we observe a galaxy or a quasar at the distance x and at the time $t = -x/c$, we have $\alpha = -\gamma$. Of consequence the recession law $\beta = \alpha/(1 + \gamma)$ becomes $\beta = \alpha/(1 - \alpha)$, or $1 + \beta = 1/(1 - \alpha)$. Substituting in (4.98) we have $1 + z = 1/(1 - \alpha)$, that is $z = \beta$ or $V = cz$. Thus, the PSR distance-red shift relation is:

$$x = rz/(1+z). \qquad (4.99)$$

4.9   The Meaning of Physical Quantities in PSR

Let us consider two observers O and O' and let H(O|O') be the value of the physical quantity H in the place occupied by observer O but defined in the reference frame of observer O'. Let instead H(O'|O) be the value of the same quantity in the place occupied by observer O', defined in the reference frame of observer O. Let us then indicate with H(O|O) the value of H in the place occupied by observer O, as defined in the reference frame of O, and with H(O'|O') the value of H in the place occupied by observer O', defined in the reference frame of observer O'. The quantity H can be, for example, the gravitational or the magnetic field, the speed of light in the vacuum, etc.

That which observer O can actually measure, through an interaction, is H(O|O); similarly, observer O' can measure H(O'|O'). It is essential to understand that O cannot measure H(O'|O), nor can O' measure H(O|O'), because every measurement is an event and therefore is local. However, Fantappié-Arcidiacono transformations provide values of H(O|O') starting from, say, H(O|O); or the values of H(O'|O) starting from H(O'|O'). What, therefore, is the physical meaning of H(O|O'), H(O'|O)?

One must bear in mind that the laws of propagation of physical phenomena formulated in the reference system of O give H(O'|O) as a function of H(O|O); the same laws, formulated in the system of O', connect H(O|O') to H(O'|O'). This is evident is one takes, as an example of the quantity H, a continuous field—magnetic, gravitational, etc.—though this restriction is not at all necessary. Thus, the "non measurable" quantity H(O|O') is related to the directly measurable quantity H(O'|O') through the laws of propagation; but H(O|O') can in turn be linked to the directly measurable quantity H(O|O) through Fantappié-Arcidiacono transformations. Thus, there actually is a link between two directly measurable quantities, namely H(O|O) and H(O'|O').

The difference between PSR and SR is that the parameter r (or, which is the same, t_0) enters into both the passages which express this relation in PSR (law of propagation and transformation of the inertial reference frame), and therefore the causal link between distant events is affected by the global curvature of spacetime. Obviously, local interaction processes, i.e. those which involve energy exchanges over small distances compared to r or over brief times compared to t_0, are not affected by the curvature. Therefore, physical quantities such as the dimension of bounded states (atoms, galaxies, etc.), the energy levels of bounded states, and so on, do not show any variation in PSR, whereas the link by means of signals between distant events does. For example, there will be a difference between the frequency of a light wave emitted by a galaxy, measured at the start, and the frequency of the same wave measured on its arrival in another galaxy. This is precisely what "red shift" consists of.

Though PGR has not yet been sufficiently investigated from this point of view, it is plausible that the topics discussed in this section can to a certain extent be relevant to it. The most important difference is that global reference frames associated with the observers O and O′ no longer exist: the reference frames introduced by theory are now local. On the space tangent in O at the manifold X which generalizes the De Sitter chronotope, laws of propagation similar to those of PSR are still defined, and these still connect H(O′|O) to H(O|O). Yet, the relation between the quantity H(O′|O) thus introduced and the quantity H(O′|O′), defined in the origin of the space tangent in O′ to X, is no longer expressed by global transformations such as the Fantappié-Arcidiacono ones. This relation is now expressed by the projective connection associated with the fundamental tensor of the metric which generalizes (4.5) [8, 9, 20, 25].

In the practical use of PSR it is necessary accurately to define the suitable physical quantities of a problem, because the global curvature effects associated with spacetime translations (effects which do not exist in SR) can easily lead to paradoxes. Let us consider, for example, the case in which the quantity H is the spatial position x of a material point in an isolated bounded system, and the concerned law of propagation is the equation of motion $x = x(t)$, solution of (4.35). This equation is valid in an inertial reference frame whose origin is in the observation pointevent O, and t is the chronological distance from O. In this reference frame, the system to which the material point belongs is bounded and its centre of mass is assumed to be at rest; thus, one would be tempted to define the notion of "bounded system" by asserting that $| x | < R$, where R is a constant of motion. As can easily be seen, this notion of "bounded system" is inconsistent in PSR, as it is incompatible with that of the inertial observer at rest with respect to the system. Such an observer evolves from event O to event O′, which is the origin of a reference frame in which the event (x, t) is simultaneous with O′. From the general expression of coordinate transformations for time translations (1-dimensional case) one has [17–21]:

$$x' = \frac{x\sqrt{1-\gamma^2}}{1+\gamma\frac{t}{t_0}}; \quad t' = \frac{t+T_0}{1+\gamma\frac{t}{t_0}}, \qquad (4.100)$$

where $\gamma = T_0/t_0$. It follows, assuming the request for simultaneity $t' = 0$, that $T_0 = -t$ and:

$$x' = \frac{x}{\sqrt{1-\left(\frac{t}{t_0}\right)^2}}. \qquad (4.101)$$

One clearly sees that in the translated reference frame the spatial position of the material point diverges for $t \to \pm t_0$, and therefore the notion of "bounded system" introduced above cannot be exported to the new reference frame. Physically, however, the observer is causally disconnected from events external to his lightcone, and therefore the divergence expressed by (4.101) does not have any consequences on how he sees the bounded system. The mistake consists in having introduced a notion of "bounded system" using a spatial position external to the observer's lightcone. This mistake can be remedied by introducing a different notion, which uses quantities internal to the lightcone. For example, one can say that the system is bounded in the sense that the travel of a ray of light from any of its parts to the observer have a duration not exceeding R/c, where R is a constant of motion. As can easily be seen, this definition is invariant for time translations, i.e. it does not depend on the fact that the observer coincides instantaneously with O or O'.

One must however pay attention to the fact that while the duration of the light travel from one part of the system to the observer is invariant for time translations, the duration of the light travel between a given emission pointevent and a given observation pointevent such as O is instead changed by the action of transformations (4.100).

4.10   Concluding Notes

Most works concerning PSR are available in Italian, and this fact has probably contributed to the limited dissemination of this theory among specialists. This chapter wishes to present a summary of the fundamental PSR equations that is comprehensible to a wider public and can lead on to more specialized studies. The fundamental dynamics equations of material point, perfect incompressible fluid and wave have been summarized and derived following more direct reasoning than can be found in the original texts. Some mistakes have been corrected or eliminated. For example, by generalizing Maxwell's equations in the De Sitter-invariant form on the Castelnuovo chronotope, a longitudinal component of the electromagnetic field appears, because of the finite value of r [40], which satisfies equations similar to those of perfect fluid hydrodynamics [20]. Arcidiacono was convinced, on the basis of this purely formal analogy [36–39], that the longitudinal component was the hydrodynamic field! In this chapter, his "cosmic magnetohydrodynamic" model, based on these principles, has been completely ignored.

As can be seen from the discussion on material point dynamics (Sect. 4.4) and as confirmed by the right-hand members of (4.68) and (4.85), an important difference with respect to SR is constituted by the dynamic effect of geodetic projection. In the conventional description of the De Sitter chronotope, this effect is, at least in part, recovered by introducing a "cosmological term" that does not exist in PSR. PSR thus becomes a useful model, at least to understand the possible kinematic origin of the cosmological term.

Again, on a kinematic basis, it is possible also to deduce a phenomenon of expansion of the Universe with a velocity field expressed by (4.52), which disappears in Einstein's limit $r \to \infty$. This field diverges at a chronological distance from the observer which is equal to $-t_0$, but this singularity—as has been discussed in other works [26, 41]—cannot be identified with the big bang. Rather, it is a horizon dependent upon the observer.

The projective effects do not affect interaction phenomena; these are still correctly described by SR, since they are local. For example, PSR cannot be taken as the basis to explain a cosmological variation of the fundamental constants. The projective effects, on the other hand, affect the propagation of signals between events that are distant in time and/or in space. For example, a discussion of travelling waves shows [20] that they are subject to a Doppler effect that can be related to cosmic expansion. The frequency of the light wave emitted by a galaxy and measured in the reference frame of the emitting galaxy in the place of emission differs from the frequency of the same wave on its arrival in another galaxy, measured in the reference frame of the galaxy of arrival. This is what the cosmological "red shift" predicted by PSR consists of; its origin is entirely due to the Doppler effect and not to the variation of the distance scale (whereas in PGR there is a contribution deriving from this variation, [41]). This entire topic can be generalized to any quantity H, as illustrated in Sect. 4.9 and, with reference to temperature, in Sect. 4.6.

References

1. Fantappié, L.: Rend. Accad. Lincei XVII, fasc. 5 (1954)
2. Fantappié, L.: Collectanea Mathematica XI, fasc. 2 (1959)
3. Arcidiacono, G.: Rend. Accad. Lincei XX, fasc. 4 (1956)
4. Arcidiacono, G.: Collectanea Mathematica X, 85–124 (1958)
5. Arcidiacono, G.: Collectanea Mathematica XII, 3–32 (1960)
6. Arcidiacono, G.: Collectanea Mathematica XIX, 51–72 (1968)
7. Arcidiacono, G.: Collectanea Mathematica XX, 231–256 (1969)
8. Arcidiacono, G.: Collectanea Mathematica XVI, 149–168 (1964)
9. Arcidiacono, G.: Collectanea Mathematica XXXIV, 95–107 (1964)
10. Kerner, H.E.: Proc. Natl. Acad. Sci. U.S.A. 73, 1418–1421 (1976)
11. Iovane, G., Giordano, P., Laserra, E.: Chaos Solitons Fractals 22(5), 975–983 (2004) arXiv: math-ph/0405056v1
12. Aldrovandi, R., Bertrán Almeida, J.P., Pereira, J.G.: (2007) arXiv:gr-qc/0606122v2
13. Cacciatori, S., Gorini, V., Kamenshchik, A.: Ann. Der Physik 17, 728–768 (2008)
14. Aldrovandi, R., Pereira, J.G.: Found. Phys. 39, 1–19 (2009)
15. Han-Ying, G., Chao-Guang, H., Zhan, X., Bin, Z.: (2004) arXiv:hep-th/043171v1
16. Han-Ying, G.: Phys. Lett. B 653(1), 88–94 (2007)
17. Arcidiacono, G.: Projective Relativity. Cosmology and Gravitation. Hadronic Press, Nonantum (USA) (1986)
18. Arcidiacono, G.: The Theory of Hyper-spherical Universes. International Center for Comparison and Synthesis, Rome (1987)
19. Arcidiacono, G.: Hadron. J. 16, 277–285 (1993)
20. Arcidiacono, G.: La teoria degli universi, vol. I-II. Di Renzo, Rome (2000) (in Italian)
21. Licata, I.: El. J. Theor. Phys. 10, 211–224 (2006)
22. Castelnuovo, G.: Rend. Accad. Lincei XII, 263 (1930)
23. Castelnuovo, G.: Scientia 40, 409 (1931)
24. Castelnuovo, G.: Mon. Not. Roy. Astron. Soc. 91, 829 (1931)
25. Pessa, E.: Collectanea Mathematica XXIV, 151–174 (1973)
26. Licata, I., Chiatti, L.: Int. J. Theor. Phys. 48(4), 1003–1018 (2009)
27. Bondi, H.: Cosmology. Cambridge University Press, Cambridge (1961)

28. Arcidiacono, G.: Collectanea Mathematica XXIII, 105–128 (1972)
29. Arcidiacono, G.: Collectanea Mathematica XXVI, 39–66 (1975)
30. Ott, H.: Zeits. Phys. **175**, 70 (1963)
31. Touschek, B., Rossi, G.: Meccanica Statistica. Boringhieri, Torino (1970) (in Italian)
32. Hakim, R., Mangeney, A.: Nuovo Cimento Lett. I (9), 429–435 (1969)
33. Arcidiacono, G., Capelas de Oliveira, E.: Hadron. J. 14, 353 (1991)
34. Arcidiacono, G., Capelas de Oliveira, E.: Hadron. J. 14, 137 (1991)
35. Gomes, D.: O potential generalisado no Universo de De Sitter-Castelnuovo. Thesis, Campinas (Brazil) (1994) (in Portuguese)
36. Arcidiacono, G.: Rend. Accad. Lincei XVIII, fasc. 4 (1955)
37. Arcidiacono, G.: Rend. Accad. Lincei XVIII, fasc. 5 (1955)
38. Arcidiacono, G.: Rend. Accad. Lincei XVIII, fasc. 6 (1955)
39. Arcidiacono, G.: Rend. Accad. Lincei XX, fasc. 5 (1956)
40. Roman, P., Aghassi, J.J.: J. Math. Phys. **7**, 1273 (1966)
41. Chiatti, L.: El. J. Theor. Phys. 15(4), 17–36 (2007) arXiv:physics/0702178

Chapter 5
Cosmic Electromagnetism

Electromagnetism is written in invariant form for the de Sitter group and analyze
the role of 5-dimensional quantities.

5.1 Electromagnetic Equations in PSR

In their conventional form, the Maxwell equations of electromagnetism and the
expression of the Lorentz force acting on charges and currents are covariant with
respect to the Poincaré group. In PSR these relations must be appropriately generalized so as to be covariant with respect to the De Sitter group.

In the usual three-dimensional space (when the time dimension is neglected) the
Maxwell equations reduce to the electrostatic equations:

$$\begin{cases} Div\, E_k = \rho \\ Curl\, E_k = 0 \end{cases} \tag{5.1}$$

where the operators Div and $Curl$ are suitably defined on this space and E is the
ordinary electric field. In the Minkowski $(3 + 1)$-dimensional space the Maxwell
equations take the usual form:

$$\begin{cases} Div\, f_{ik} = j_k \\ Curl\, f_{ik} = 0 \end{cases} \tag{5.2}$$

where f_{ik} is the electromagnetic tensor and j is the current density. The operators Div
and $Curl$ are now defined on Minkowski space.

© The Author(s) 2017   67
I. Licata et al., *De Sitter Projective Relativity*, SpringerBriefs in Physics,
DOI 10.1007/978-3-319-52271-5_5

Equations (5.1) and (5.2) can be expressed by the same form [1]:

$$\begin{cases} Div\, G_{ik} = 0 \\ Curl\, G_{ik} = j_{ikl} \end{cases} \tag{5.3}$$

provided that:

$$G_{ik} = \begin{pmatrix} 0 & E_3 & E_2 \\ -E_3 & 0 & E_1 \\ E_2 & -E_1 & 0 \end{pmatrix} \tag{5.4}$$

in the electrostatic case and

$$G_{ik} = \begin{pmatrix} 0 & E_3 & E_2 & iH_1 \\ -E_3 & 0 & E_1 & iH_2 \\ E_2 & -E_1 & 0 & iH_3 \\ -iH_1 & -iH_2 & -iH_3 & 0 \end{pmatrix} \tag{5.5}$$

in the electromagnetic case. The H vector is the usual magnetic field. Moreover, the following integrability condition is requested:

$$Curl\, j_{ikl} = 0. \tag{5.6}$$

Equations (5.3) and (5.6) do not depend on space dimensionality. This property is shared by the ponderomotive force, when expressed in the form:

$$F_l = \frac{1}{2} G^{ik}\, j_{ikl}. \tag{5.7}$$

Is then possible to apply (5.3), (5.6) and (5.7) to the de Sitter spacetime, thus obtaining [1]:

$$G_{ik} = \begin{pmatrix} 0 & E_3 & E_2 & iH_1 & -C_1 \\ -E_3 & 0 & E_1 & iH_2 & -C_2 \\ E_2 & -E_1 & 0 & iH_3 & -C_3 \\ -iH_1 & -iH_2 & -iH_3 & 0 & -C_0 \\ C_1 & C_2 & C_3 & C_0 & 0 \end{pmatrix} \tag{5.8}$$

Such a generalization gives the following expressions, where the notations have the usual meaning [2–4]:

$$\begin{aligned} div\, E &= \rho \\ div\, H + \bar{\partial}_5\, C_0 &= 0 \\ curl\, E + \bar{\partial}_0\, H - \bar{\partial}_5\, C &= 0 \end{aligned} \tag{5.9}$$

$$curl\,\boldsymbol{H} - \bar{\partial}_0\,\boldsymbol{E} = \boldsymbol{j}$$
$$div\,\boldsymbol{C} + \bar{\partial}_0\,C_0 = s$$
$$curl\,\boldsymbol{C} - \vec{\partial}_5\,\boldsymbol{E} = 2\boldsymbol{\omega} \tag{5.10}$$
$$grad\,C_0 + \bar{\partial}_0\,\boldsymbol{C} - \bar{\partial}_5\,\boldsymbol{H} = \boldsymbol{a}$$

The expression of ponderomotive force is as follows:

$$\boldsymbol{F} = (\rho\,\boldsymbol{E} + \boldsymbol{j} \times \boldsymbol{H}) - (\boldsymbol{a}\,C_0 + 2\,\boldsymbol{\omega} \times \boldsymbol{C})$$
$$F_0 = i(\boldsymbol{j} \cdot \boldsymbol{E} - \boldsymbol{a} \cdot \boldsymbol{C}) \tag{5.11}$$
$$F_5 = \boldsymbol{a} \cdot \boldsymbol{H} - 2\boldsymbol{\omega} \cdot \boldsymbol{E}$$

These results require a comment. As one can see, in addition to the usual electric $\boldsymbol{E}$ and magnetic $\boldsymbol{H}$ fields appear two new fields: one a scalar (C_0), and the other a vector ($\boldsymbol{C}$). In the corresponding generalization of the ponderomotive force (5.11), these fields are coupled with the currents $\boldsymbol{a}$ ed ω and the charge density s. For a correct reading of the equations, there should be kept in mind that the derivatives that appear are projective derivatives, and not conventional partial derivatives.

Equations (5.9) take the name of "Maxwellian" equations, while (5.10) are known as "non-Maxwellian" equations. It is not difficult to understand the genesis of this second group of equations. In an infinite universe, the photon has a null rest mass M and it moves at the speed of light; its Compton wavelength is thus $\lambda_C = \hbar/Mc = \infty$. But if the radius of the universe r is finite, we cannot have $\lambda_C >> r$ and thus $\lambda_C = \lambda_{C\,max} \approx r$. It follows that in a certain sense the photon acquires a finite mass $\approx \hbar/rc$. It therefore acquires a longitudinal component and this is the reason for the appearance of the fields ($\boldsymbol{C}, C_0$). Equations (5.10) formally coincide, for $r = \infty$, with the non-Maxwellian photon equations obtained by de Broglie with the fusion method on the Minkowski spacetime (when the rest photon mass is negligible); these latter equations, in fact, describe precisely the longitudinal component of the photon [5, 6].

For $r \rightarrow \infty$, (5.9) and (5.10) are not more coupled and we must recover the ordinary Maxwell equations that were the starting point of the generalization; therefore the disappearance of non-Maxwellian equations is requested. In turn, the necessary disappearance of the non-Maxwellian fields implies the inexistence of distinct sources of these fields; in other words, there must be identically $s, a, \omega = 0$. The only origin of the fields ($\boldsymbol{C}, C_0$) is therefore the relativistic transformation (in the framework of PSR) of the ordinary fields ($\boldsymbol{E}, \boldsymbol{H}$) associated with remote charges and currents. Let's consider, as an example, the equations:

$$C_0' = C_0;$$
$$C' = (C - \gamma H) / (1 - \gamma^2)^{1/2};$$
$$E' = E;$$
$$H' = (H + \gamma C) / (1 - \gamma^2)^{1/2}.$$

(5.12)

which express the transformation of the fields due to a time translation of parameter T_0. We have set $\gamma = T_0/t_0$, where $t_0 = r/c$ and r is the radius of curvature of the de Sitter spacetime. It is immediately evident that at the Minkowskian limit these relations become invariances. Given the absence of local sources of non-Maxwellian fields, these latter must be null here and now. Therefore $C_0 = 0$, $C = 0$ and:

$$C_0' = 0;$$
$$C' = -\gamma H / (1 - \gamma^2)^{1/2};$$
$$E' = E;$$
$$H' = H / (1 - \gamma^2)^{1/2}.$$

(5.13)

The second relation expresses the appearance of a field C' as transformation of the magnetic field H at cosmological distances. Thus, if we observe the effects of a very distant magnetic field in the past (let's say five billion years ago) these effects will be modified due to an increase in the field strength by a factor of $1/(1 - \gamma^2)^{1/2}$, and due to the simultaneous appearance of a field C' associated with the transformation of H.

How we can see by substituting the identities $s = 0$, $a = 0$, $\omega = 0$[1] into (5.11), the fields (C, C_0) are not coupled with the matter and thus, in particular, they do not contribute to the luminosity of a remote astronomical object. More generally, the energy associated with the non-Maxwellian fields will appear, in the energy balance of the electromagnetic field, as "missing" energy which is not instrumentally detectable in a direct manner.

Obviously, if the relativistic transformations of the fields E ed H that alter the luminosity of extragalactic objects are not taken into account, the evaluations of the extragalactic scale distance using standard candles can be affected by significant

---

[1]These equations are the duals of the relations expressing the inexistence of the magnetic charge and magnetic current. In fact, the magnetic field itself derives from a relativistic transformation of the electric field in the context of conventional SR: it is generated by electric fields *in motion*, like the fields C derive instead from the *distant* magnetic fields. All this is due to the simultaneous existence in PSR of two fundamental constants: the *speed c* and the *distance r*.

systematic errors. A study of the problem requires the evaluation of how the Poynting vector transforms passing from a reference frame with origin at the remote source to a new reference frame with origin at the observation pointevent.

We remark that PSR is a five-dimensional theory, hence we have two Poynting vectors [2–4]:

$$\begin{aligned} \boldsymbol{T}_{\alpha 0} &= i(C_0\,\boldsymbol{C} + \boldsymbol{E} \times \boldsymbol{H}) \\ \boldsymbol{T}_{\alpha 5} &= C_0\,\boldsymbol{H} + \boldsymbol{E} \times \boldsymbol{C} \end{aligned} \tag{5.14}$$

The physical interpretation of these vectors is not difficult: their fluxes through a closed hypersurface containing the entire three-dimensional space provides the components of the momentum of the electromagnetic field along the time axis and along the fifth axis respectively. The component of the momentum along the fifth axis does not have direct physical effects, so only the first of (5.14) is relevant. This equation generalizes the ordinary Poynting vector including the flux of energy associated with the longitudinal component of the electromagnetic waves. In the Einsteinian limit $r \to \infty$ the first of (5.14) reduces to the ordinary Poynting vector of normal electromagnetic theory, while the second equation transforms into the identity $0 = 0$.

5.2 Relativistic Transformation of Electromagnetic Quantities (in the Framework of PSR)

The transformation of coordinates connecting the two reference frames consists in the product of a time translation of the origin with parameter T_0, a space translation of the origin with parameter T and a boost with speed V. We will set, following the customary notation in the literature, $\alpha = T/r$, $\beta = V/c$, $\gamma = T_0/t_0$.

The general transformation of homogeneous coordinates takes the form:

$$\bar{x}_A = a_{AB}\,\bar{x}_B \tag{5.15}$$

where A, B = 0, 1, 2, 3, 5. Recalling that (Chap. 2)

$$x' = \frac{Ax + [\beta + \gamma(\alpha - \beta\gamma)]ct + BT}{A(\gamma\beta - \alpha)\,x/r + (\gamma - \alpha\beta)\,t/t_0 + B} \tag{5.16}$$

$$t' = \frac{A\,\beta x/c + [1 + \alpha(\alpha - \beta\gamma)]\,t + BT_0}{A\,(\gamma\beta - \alpha)\,x/r + (\gamma - \alpha\beta)\,t/t_0 + B} \tag{5.17}$$

we have

$$B\,a_{11} = 1$$
$$B\,a_{01} = \beta$$
$$B\,a_{51} = \beta\gamma - \alpha$$
$$AB\,a_{10} = \beta + (\alpha-\beta\gamma)\gamma$$
$$AB\,a_{00} = 1 + (\alpha-\beta\gamma)\alpha$$
$$AB\,a_{50} = \gamma - \alpha\beta$$
$$A\,a_{15} = \alpha$$
$$A\,a_{05} = \gamma$$
$$A\,a_{55} = 1$$
$$A^2 = 1+\alpha^2-\gamma^2$$
$$B^2 = 1-\beta^2 + (\alpha-\beta\gamma)^2;$$
$$a_{12} = a_{13} = a_{21} = a_{23} = a_{20} = a_{25} = a_{31} = a_{32} = a_{30} = a_{35} = a_{02} = a_{03} = a_{52} = a_{53} = 0;$$
$$a_{22} = a_{33} = 1.$$

$$(5.18)$$

The electromagnetic field is expressed by the tensor H_{AB}, dual of G_{AB}:

$$H_{kl} = E_k; \quad H_{k0} = -iH_k; \quad H_{k5} = C_k; \quad H_{05} = iC_0; \quad k,l = 1,2,3. \quad (5.19)$$

The general transformation rules of the electromagnetic field components are then:

$$H'_{AB} = a_{AR}\,a_{BS}\,H_{RS}; \quad R,S = 0,\,1,\,2,\,3,\,5 \qquad (5.20)$$

that is [7]:

$$E'_1 = E_1$$
$$E'_2 = a_{11}E_2 + a_{10}H_3 + a_{15}C_3$$
$$E'_3 = a_{11}E_3 + a_{10}H_2 + a_{15}C_2$$
$$H'_1 = a_{51}C_0 + a_{50}C_1 + a_{55}H_1$$
$$H'_2 = a_{01}E_3 + a_{00}H_2 + a_{05}C_2$$
$$H'_3 = a_{01}E_2 + a_{00}H_3 + a_{05}C_3 \qquad (5.21)$$
$$C'_0 = a_{11}C_0 + a_{10}C_1 + a_{15}H_1$$
$$C'_1 = a_{01}C_0 + a_{00}C_1 + a_{05}H_1$$
$$C'_2 = a_{51}E_3 + a_{50}H_2 + a_{55}C_2$$
$$C'_3 = a_{51}E_2 + a_{50}H_3 + a_{55}C_3.$$

References

1. Arcidiacono G.: *Collectanea Mathematica* X, 85–124 (1958)
2. Arcidiacono, G.: Projective Relativity. Cosmology and Gravitation. Hadronic Press, Palm Harbour (1986)
3. Arcidiacono, G.: The theory of hyperspherical universes. International Center for Comparison and Synthesis, Rome (1987)
4. Arcidiacono, G.: La teoria degli universi, vol. II. Di Renzo, Rome (2000)
5. De Broglie L.: Théorie générale des particules a spin. Gauthier-Villars, Paris, 1954
6. De Broglie, L.: Une nouvelle conception de la lumière. Hermann, Paris (1954)
7. Arcidiacono G.: *Collectanea Mathematica* XXV, 159–184 (1974)

Chapter 6
Projective General Relativity (PGR)

Einstein equations are generalized for de Sitter 5-dimensional space, and their local projections (PGR), with some important cosmological considerations, are studied.

6.1 From PSR to PGR

The theory presented in the preceding chapters is based on the de Sitter group as a *global* group of physical law symmetry, and is therefore equivalent to Einstein's Special Relativity (SR), in which the same function is performed by the Poincaré group. This theory, given the importance of projective representation in its formulation, was called Projective Special Relativity (PSR) by Arcidiacono [1], who was the first to formulate it.

The essential difference between PSR and ordinary SR lies in the presence of the de Sitter horizon. Compared with SR, this presence does not lead to modifications in the local physics, i.e., in the spacetime neighbourhood of the observer; it does produce, however, projective distortions of the results of measurements made by that observer on processes remote from him. The adjective remote must be understood here in respect of both space and time; if x and t are the spatial and time distances, respectively, of a given event from the observer and r and $t_0 = r/c$ the de Sitter radius and time, respectively, the event is remote if x/r or t/t_0 are not negligible. The most evident distortions are the *purely kinematic* red shift of a light signal emitted by an extragalactic source and the dependence of the light cone aperture on the spatial and time distance from the observer [2]. PSR therefore has cosmological aspects that are absent in SR and one may wonder whether it constitutes a plausible cosmological model.

It is not difficult, however, to realize that PSR cannot constitute a logically consistent cosmological model. Its metric does not describe gravitation, and to attempt a non-metric description of gravitation through a de Sitter-invariant field theory that generalizes the Poincaré-Nordström one has the drawback, already

I. Licata et al., *De Sitter Projective Relativity*, SpringerBriefs in Physics,
DOI 10.1007/978-3-319-52271-5_6

known in SR, of incorrect predictions of light deflection. But, even if the problem of describing gravitation is set aside, serious problems arise as to logical consistency. The crucial point is the absence of a cosmic time: in PSR every observer coordinates events using his own local clock and there is no defined synchronization agreement among the time measurements of the various observers other than that which is implicit in the PSR transformations of the coordinates. Indeed, if such an agreement did exist, it would clash with such transformations, i.e., it would destroy the de Sitter invariance.

The non-existence of a cosmic time implies the non-validity of the cosmological principle as it is normally understood. PSR kinematics might conceivably be compatible with the *perfect* cosmological principle, i.e., with a steady-state cosmology. Unfortunately, though, this solution, too, leads to a new self-contradiction, irrespective of the fact that steady-state cosmology is no longer supported by observations. In PSR, the extension of the time domain marked by each local clock is infinite and this, in the steady-state hypothesis, would mean that matter has been "ageing" for an infinite time. Thus the Sun and other stars should no longer be shining as they were exhausted an infinite time ago. They could not be gradually replaced by other stars such as to ensure that a steady-state equilibrium is maintained, since in PSR a matter creation field similar to that postulated in the old Bondi, Gold and Hoyle steady-state theory does not exist. Any attempt to introduce this term "by hand" would violate the invariance of physical laws respect to time translations (and therefore the de Sitter invariance), at the level where the supposed creation of matter, say atomic level, would occur. In conclusion, therefore, the only way to satisfy the perfect cosmological principle whilst preserving the de Sitter invariance would be to postulate a Universe that is entirely and perpetually empty. The vacuum does not have problems with regard to ageing. PSR therefore must constitute the empty solution of a more general theory that comprises matter and gravitation. That is, a Projective General Relativity (PGR) must exist that includes PSR as its empty solution, just as ordinary General Relativity (GR) comprises SR as the "empty" limiting case. This conclusion leads to believe that the formal relation between PGR and PSR is the same as that between GR and SR. This reflection led Arcidiacono to the first actual construction of such a theory in 1964 [3], and it will be set forth in Sect. 6.2.

The construction of PGR was followed by applications on a local scale, to which a brief reference will be made in Sect. 6.3.

The definition of the cosmological problem within the PGR context requires additional specifications with respect to Fridman's ordinary cosmology within the GR context. In both cases, solutions must be identified that satisfy the (ordinary) cosmological principle. This principle leads to the existence of a scale distance that is a function of cosmic time and this result is not compatible with the de Sitter group, which thus can no longer be valid as a global invariance group. Unlike the case of Fridman cosmology, however, there must also exist a generalization of the Cayley-Klein absolute and a parameter r that generalizes the de Sitter PSR radius. For $r \to \infty$ the corresponding Fridman cosmological model must result. The two points that distinguish the cosmological problem in the PGR version from that in the GR version are, therefore, the presence of a (generalized) de Sitter horizon and the ensuing fact that a part of the recessional motion of galaxies from the observer is not the result of spatial expansion but of the kinematics connected with the existence of this horizon. In the $r \rightarrow \infty$ limit, the chronological distance of the de Sitter horizon from the generic observer increases indefinitely and the kinematic component of the recession vanishes; the ordinary spatial expansion/contraction foreseen by GR remains. The presence of a kinematic factor in the variation of the scale distance in PGR leads to a precise relation between the PGR cosmic time scale and the cosmic time scale of the corresponding Fridman GR model. For large cosmic time values (i.e., at a great chronological distance from the initial singularity), this relation is expressed by Milne's double time scale, as can be seen by a simple argument based on an appropriate modification of Newtonian cosmology by Milne and McCrea (Sect. 6.5). More generally, in a model with an initial singularity and indefinite expansion the observer sees the singularity approaching the horizon asymptotically as cosmic time increases. The relation between cosmic times is thus expressed by a generalized Milne scale, and this fact is equivalent to the appearance of a cosmological term (Sect. 6.6). The PGR approach therefore leads to a natural genesis of the cosmological term and of "dark energy". As we shall see, it also throws new light on issues such as the origin of inertia (Sect. 6.8) and the cosmological background of dark matter (Sect. 6.9).

6.2   Arcidiacono's Construction

The first definition of PGR can be found in a 1964 article by Arcidiacono [3]. In this paper, the de Sitter PSR hyperboloid, immersed in a five-dimensional space, is generalised in the form of a four-dimensional manifold X_4, in each point of which a system of local curvilinear coordinates (y^μ, y^5) is defined, with $\mu = 0, 1, 2, 3$. On the space tangent to the manifold in this point, a system $(\underline{x}^\mu, \underline{x}^5)$ of homogeneous projective coordinates is then defined, again with $\mu = 0, 1, 2, 3$. The transport law is defined by:

$$d\underline{x}^A + \omega_B^A \underline{x}^B = 0, \tag{6.1}$$

where:

$$\omega_B^A = \pi_{BC}^A \, d\underline{x}^C \tag{6.2}$$

and $A, B, C = 0, 1, 2, 3, 5$. On the same tangent space a quadric is then defined:

$$\gamma_{AB} \underline{x}^A \underline{x}^B = 0 \tag{6.3}$$

which defines a local light cone. The projective connection expressed by the π coefficients must define a projective translation law that leaves the quadrics field (6.3) unchanged. That is, it must be:

$$\nabla_A \, \gamma_{BC} = 0. \tag{6.4}$$

This condition is satisfied by Veblen's projective connection:

$$\pi_{BC}^A = \frac{1}{2}\gamma^{AS} \left(\underline{\partial}_C \, \gamma_{BS} + \underline{\partial}_B \, \gamma_{CS} - \underline{\partial}_S \, \gamma_{BC} \right) \tag{6.5}$$

in which the indices run along 0, 1, 2, 3, 5 and $\underline{\partial}_A = \partial/\partial \underline{x}_A$. Starting from this connection, the projective curvature tensor is obtained:

$$R_{BCD}^A = \underline{\partial}_C \, \pi_{BD}^A - \underline{\partial}_D \, \pi_{BC}^A + \pi_{SC}^A \, \pi_{BD}^S - \pi_{SD}^A \, \pi_{BC}^S. \tag{6.6}$$

This tensor is null in a projectively flat space, i.e., in a space having *constant curvature*. Despite of its name, tensor (6.6) actually also describes the torsion [4]. This can be seen by examining the structure equations on X_4:

$$\Omega_B^A = d\omega_B^A + \omega_S^A \wedge \omega_B^S; \tag{6.7}$$

the Ω_5^μ components, with $\mu = 0, 1, 2, 3$ define the torsion. The gravity equations thus take the form:

$$G_{AB} = R_{AB} - \frac{1}{2} R \gamma_{AB} = \chi \, T_{AB} \tag{6.8}$$

where the meaning of the terms is the conventional one.

One can note that as regards the definition of the quadrics field and of the connection, the construction is very similar to that proposed by Veblen in his Projective Relativity [5]; nevertheless, there are also important differences. In Veblen's theory, the metric coefficients $\gamma_{\mu5}$ are identified with the components of the electromagnetic potential quadrivector, and they must therefore be independent of $\underline{x}_5$, while no restriction of this type exists in the construction presented here. Veblen's aim was to reach a formal unification of Maxwell's equations and of Einstein's gravity equations as different components of (6.8). Thus Veblen's theory does not envisage an actual 5-dimensional generalization of gravitational phenomena, which, instead, is the goal of Arcidiacono's investigation [6].

The projective metric is expressed by the equation:

$$ds^2 = \gamma_{AB} \, d\underline{x}^A \, d\underline{x}^B, \tag{6.9}$$

And the projective coordinates are normalised in accordance with the condition:

$$\gamma_{AB}\, \underline{x}^A \underline{x}^B \;=\; r^2 \tag{6.10}$$

which generalizes the PSR Cayley-Klein absolute. The relation between the projective coordinates and the physical ones is expressed by the relations:

$$\underline{x}_\mu = x_\mu/A \tag{6.11a}$$

$$\underline{x}_5 = r/A \tag{6.11b}$$

$$A^2 = \gamma_{55} + 2\gamma_{\mu 5}\,\frac{x^\mu}{r} + \gamma_{\mu\nu}\,\frac{x^\mu x^\nu}{r^2}. \tag{6.11c}$$

The projective derivatives can thus be expressed as a function of the ordinary partial derivatives through the relations:

$$\overline{\partial}_\mu = A\,\partial_\mu \tag{6.12a}$$

$$\overline{\partial}_5 = -\frac{A}{r}x_\mu\,\partial^\mu. \tag{6.12b}$$

Because of Eqs. (6.11a, 6.11b, 6.11c), the projective metric (6.9) induces a metric on the physical coordinates:

$$ds^2 = g_{\mu\nu}\,dx^\mu\,dx^\nu, \tag{6.13}$$

in which the Greek indices run along 0, 1, 2, 3 and:

$$g_{\mu\nu} = A^{-4}\left[A^2\,\gamma_{\mu\nu} + (Y_\mu - X_\mu)\,(Y_\nu - X_\nu)\right] \tag{6.14}$$

$$X_\mu = \gamma_{\mu 5} + \gamma_{\mu\nu}\,x^\nu \tag{6.15}$$

$$Y_\mu = \frac{1}{2}\left[\partial_\mu \gamma_{55} + x^\nu \partial_\mu \gamma_{\nu 5} + x^\xi x^\nu\,\partial_\mu \gamma_{\xi\nu}\right]. \tag{6.16}$$

Equation (6.14) is therefore the expression of the metric coefficients in physical coordinates. They can be broken down into a symmetrical part:

$$g_{(\mu\nu)} = A^{-4}\left(A^2\,\gamma_{\mu\nu} - X_\mu X_\nu + Y_\mu Y_\nu\right) \tag{6.17a}$$

and into an antisymmetrical one:

$$g_{[\mu\nu]} \;=\; A^{-4}\left(X_\mu Y_\nu - X_\nu Y_\mu\right). \tag{6.17b}$$

The induced metric, therefore, is not generally symmetrical.

It must be noted that in the local limit $x^\mu \to 0$ (i.e., in the spatial and temporal neighbourhood of the generic observer) Eqs. (6.8) are not reduced to the ordinary Einstein equations. In this limit, (6.11c), (6.15) and (6.16) become:

$$A^2 = \gamma_{55}; \quad X_\mu = \gamma_{\mu 5}; \quad 2Y_\mu = \partial_\mu \gamma_{55}. \tag{6.18}$$

If $\gamma_{55} = \phi^2$, $\gamma_{\mu 5} = \phi_\mu$, $\gamma_{\mu\nu} = a_{\mu\nu}$, these relations become:

$$A = \varphi; \quad X_\mu = \varphi_\mu; \quad Y_\mu = \varphi\psi_\mu, \quad \psi_\mu = \partial_\mu\varphi. \tag{6.19}$$

The induced metric (6.17a, 6.17b) thus takes the form:

$$g_{(\mu\nu)} = \varphi^{-2}\left(a_{\mu\nu} - \varphi_\mu\varphi_\nu + \psi_\mu\psi_\nu\right) \tag{6.20a}$$

$$g_{[\mu\nu]} = \varphi^{-3}\left(\varphi_\mu\psi_\nu - \varphi_\nu\psi_\mu\right). \tag{6.20b}$$

The antisymmetric part of the metric is connected with the vector product of vectors ϕ_μ and ψ_μ, so that it becomes null in three important cases: $\phi_\mu = 0$, $\psi_\mu = 0$ and $\phi_\mu = k\phi\psi_\mu$. In the first case, only the projective metric coefficients γ_{55} and $\gamma_{\mu\nu}$ survive, and one therefore has a *scalar-tensorial* theory. In the second case, only $\gamma_{\mu 5}$ and $\gamma_{\mu\nu}$ survive, as γ_{55} is constant (and can therefore be let to be equal to 1); one therefore has a *vector-tensorial* theory. In the third case, all the coefficients survive and one has a full *scalar-vector-tensorial* theory. General relativity corresponds to the special case $\phi = 0$, $\phi_\mu = 0$.

We cannot examine the various cases in detail, and the interested reader may refer to the original articles [6–9]. We point out, however, that in the vector-tensorial case the metric is reduced to the form (for $\phi = 1$):

$$g_{(\mu\nu)} = a_{\mu\nu} - \varphi_\mu\varphi_\nu \tag{6.21}$$

which is the same used by Veblen for his construction. Here, though, the ϕ_μ quadrivector represents a new form of gravitation, and not the electromagnetic field.

6.3   Local Applications of PGR

With consequential, but long and laborious calculations [6, 8], the field equations of the scalar-tensorial theory (in the local limit) take the following form, for a homogeneous projective metric of degree n[1]:

---

[1]Bearing in mind that a function $\varphi(x)$ is homogeneous of degree n in the x coordinates if $\varphi(\lambda x) = \lambda^n \varphi(x)$.

$$\hat{R}_{\mu\nu} - \frac{1}{2}\hat{R}a_{\mu\nu} + (3n+1)\varphi^{-1}\left(\partial_\mu\psi_\nu - a_{\mu\nu}\nabla^2\varphi\right) +$$
$$- 3n\varphi^{-2}\left[(n+1)\psi_\mu\psi_\nu + n\psi^2 a_{\mu\nu} + (n-1)\frac{a_{\mu\nu}}{r^2}\right] = \chi\frac{T_{\mu\nu}}{\varphi^2} \tag{6.22a}$$

$$(n+1)(3n-1)\partial_\mu\varphi^2 = -2\chi r T_{\mu 5} \tag{6.22b}$$

$$\hat{R} + 6n\left[\frac{\nabla^2\varphi}{\varphi} + (n-1)\frac{\psi^2}{\varphi^2} + \frac{2n}{r^2\varphi^2}\right] = -2\chi\frac{T_{55}}{\varphi^4}. \tag{6.22c}$$

In these equations, the values with the hat are calculated with respect to the $a_{\mu\nu}$ metric. In a vacuum and for $n = 0$ these equations are reduced to the following:

$$\hat{R}_{\mu\nu} + \frac{\varphi_{;\mu\nu}}{\varphi} = 0 \tag{6.23a}$$

$$\nabla^2\varphi \equiv a^{\mu\nu}\varphi_{;\mu\nu} = 0. \tag{6.23b}$$

Given a solution to (6.22a, 6.22b, 6.22c) for $n = 0$ it is possible to obtain solutions for any n through the scaling rules:

$$\gamma_{\mu\nu} = \varphi^{2n}a_{\mu\nu}; \quad \gamma_{\mu 5} = 0; \quad \gamma_{55} = \varphi^{2(n+1)}. \tag{6.24}$$

Singh and Singh [10] started from the Einstein-Rosen cylindrically symmetrical non-static metric to substitute the related components of the Ricci tensor in (6.23a, 6.23b) and thus resolve the corresponding scalar-tensorial problem in PGR. In other works they also discuss in greater depth the applications to the Liouville metric-[11] and the static case. A discussion of these contributions however is beyond the scope of this introductory presentation.

6.4   Consistency Issues

Arcidiacono's construction leads, in the local limit, to an interesting unification of aspects normally associated with generalized gravitation theories of different kinds: non-symmetrical metric, torsion, natural introduction of a fifth spacetime coordinate (clearly defined in terms of the intrinsic geometry of the customary four-dimensional spacetime), consequent possibility of TeVeS solutions. It can therefore be considered as a structure of notable physico-mathematical interest [6]. However, its full application outside of the local limit immediately leads to logical contradictions.[2] For example, these contradictions prevent its use in cosmology [12], i.e., in the very sector in which one would expect new physics from the r parameter.

The origin of these contradictions is evident and deep, and lies in the nature of the frames of reference in PSR [13, 14]. As can be recalled from the preceding chapters, such frames are *global*. A physical quantity $A(x)$ defined in the point-event x of the frame associated with an observer O is quite different from that same quantity measured by an observer O' located in x, and this is because of the projective effects induced by the finite value of the r parameter. Such effects only disappear in Einstein's $r \rightarrow \infty$ limit, in which PSR collapses onto ordinary SR. It is fully evident that the symbol $A(x)$ has no direct physical meaning, because O cannot measure $A(x)$ as he (she) is remote from x. This meaning only exists in the $x = 0$ case, i.e., at the origin of O's frame of reference, in other words in the point-event in which O is located. The symbol $A(x)$ has meaning only as a function of its appearance in the mathematical expression of the physical laws involving it, *in such a way as to ensure their de Sitter invariance*. But the real physics is given by the local expression of these laws, that is for $x = 0$.

This is an entirely general property of PSR: the presence of the de Sitter horizon does not modify the local physics, but simply introduces distortions in the propagation of signals across distances comparable to r (e.g. cosmological red shift). Such distortions are of projective origin. A good example is the PSR dispersion relation which generalises Einstein's relation between energy and impulse, by also involving the angular momentum [6, 13, 15]. In the local version of this relation, however, the angular momentum disappears and therefore in local interactions it is never coupled with linear momentum (this could only occur in a hypothetical rigid body of cosmological extension). Another example is the PSR generalization of electromagnetism, which leads to the appearance of two new components of electromagnetic field, a scalar one and a vectoral one [6, 16]. Naturally, these components are associated with the de Sitter covariance of the electromagnetic field and do not exist locally. For example, they are not coupled with electrical charges and currents. Their potential impact on experimental physics is represented by the fact that they determine a different luminosity-distance law [17].

These considerations concern PSR, but it is easily seen from (6.3) and (6.10) that the PGR frames of reference maintain a global character. Indeed, in this description every observer maintains a light cone and an absolute of his own. The considerations made in the case of PSR also continue to apply in the definition of the physical quantities in these frames. Specifically, Eq. (6.8), which establish a relationship between the local geometry and the local energy-matter distribution, can be valid *only locally*. Licata and Chiatti [12] thus proposed the following reformulation for these:

---

[2]For example, in the study of the cosmological problem the first member of (6.8) is transformed into a function of cosmic time alone, while the second member is dependent on the projective coordinates!

$$\lim_{r \to \infty} \left(R_{AB} - \frac{1}{2} \breve{R} \, \gamma_{AB} - \chi \, T_{AB} \right) = 0.$$ (6.25)

The limit operation causes the disappearance of the components with at least one of the indices equal to 5, and reduces the gravity equations to the original Einstein equations. However, the expression in parentheses admits the de Sitter group as a holonomy group, and its covariance is not therefore the ordinary covariance of the Einstein equations. The scalar and vectoral fields continue to exist as a function of this different covariance, but they disappear as actual physical fields through the limit operation. They constitute, therefore, the gravitational parallel of the additional components of electromagnetic field.

The more general expression of the projective metric is assumed to be the following:

$$ds^2 = \gamma_{\mu\nu}(\underline{x}) d\underline{x}^\mu \, d\underline{x}^\nu + d\underline{x}_5 \, d\underline{x}^5$$ (6.26)

where the $\gamma_{\mu\nu}$ coefficients are obtained by solving (6.25) and substituting the local coordinates with the projective ones in their expression. Also, the normalisation condition is assumed:

$$\gamma_{\mu\nu}(\underline{x}) \underline{x}^\mu \underline{x}^\nu + \underline{x}_5 \underline{x}^5 = r^2.$$ (6.27)

Equations (6.26) and (6.27) clearly show that the geometry-matter connection, as it constitutes a local dynamic fact, is not modified by the de Sitter horizon. Instead, the horizon modifies the kinematics in the observer's *global* frame of reference.

6.5 The Cosmological Problem

The fact that the PGR frames of reference continue to have a global structure can give rise to doubt concerning the compatibility of PGR with the cosmological principle. One can however persuade oneself of this compatibility through a non-relativistic argument, based on a suitable modification of Newtonian cosmology by Milne and McCrea [18]. It is known that this approach leads to the same set of solutions as Fridman's relativistic cosmology.

By working along these lines [17] it is possible to derive a conservation of matter equation having the form ($t_0 = r/c$):

$$\rho(t) = \frac{\rho(t_0)}{[Y(t)]^3},$$ (6.28)

in which $\rho(t)$ is the density of matter at the cosmic time instant t and the scale factor $Y(t)$ is given by the relation $Y(t) = R(t)(t/t_0)$, that is by the product of the spatial scale variation $R(t)$ by the purely kinematic (fictitious) scale variation t/t_0 resulting from the presence of the de Sitter horizon. It is thus possible to derive the Fridman equation in the form:

$$\dot{Y}^2 = \frac{8\pi G\rho(t_0)}{3Y} - k + \frac{\lambda}{3}Y^2. \qquad (6.29)$$

In (6.29) the terms have the conventional meaning and the result is that the same set of cases of (k, λ) models is obtained as in Fridman's cosmology, the only difference being that the scale distance $R(t)$ is replaced by $Y(t)$. We note, in particular, that if $R \equiv 1$, that is in the case of PSR, one obtains a empty model in which the density ρ and the cosmological term λ are both null.

One can see from the study of cosmological red shift [17] that the course of this phenomenon can be made consistent with the customary scale distance $R(t)$ if the clocks are regraduated in such a way as to measure the time τ, rather than the time t, given by:

$$\tau = t_0 + t_0 \ln\left(\frac{t}{t_0}\right), \qquad (6.30)$$

and this expression is the well-known "double time scale" introduced by Milne in his Kinematic Relativity [19]. In practice, having assigned a given Fridman cosmological model (k, λ) having cosmic time τ, one moves on to the corresponding cosmological model that is the solution to (6.29), by substituting for τ the cosmic time t given by (6.30). In the time τ (which Milne called "dynamic" or "atomic" time) there is no kinematic contribution to expansion, whereas such a contribution is present in the time t (which Milne called "kinematic" time).

Naturally, this model has an exclusively euristic value and does not constitute an actual theory; however, it does provide useful indications for the formulation of a PGR based on the cosmological principle. Specifically, by examining (6.30) one sees that while t goes from 0 to ∞, τ goes from $-\infty$ to $+\infty$; in other words, in the "equivalent" Fridman model the origin of cosmic time is situated in the infinite past. This paradox originates simply from the fact that the chronological distance from the observer, equal to $-t_0$, i.e., his de Sitter horizon, has been chosen as the origin $t = 0$ of t; but this is reasonably correct only for observers who are chronologically very far from the big bang, who therefore see the initial singularity as "flattened" on the horizon. Equation (6.30), therefore, cannot be correct for small values of t/t_0. This observation enables to understand that the truly crucial problem is actually that of a correct definition of the relationship between the (generalized) de Sitter horizon and the initial singularity. Specifically, the invariance of r with respect to the choice of the observer, which is a legacy of the PSR expressed by (6.10), implies that also observers placed on the initial singularity or in its immediate temporal vicinity must "see" a horizon at the distance t_0. But this means that the big bang cannot be the origin of space as in Fridman's cosmology. Indeed, the horizon must be pre-existent to the big bang or appear simultaneously with it; and in either case the existence must be admitted of a global geometric structure already present at the time of the big bang, and of whose context this event is part. And since the naïve model of the big bang as an explosion which takes place in an already existing spacetime leads to well-known paradoxes, the only possible alternative is that the initial singularity be simultaneous with a change of geometry. This change must consist in the transition from an "atemporal" geometry to a geometry with a defined cosmic time; the appearance of the de Sitter horizon must be the instantaneous effect of such a transition. These requirements nearly univocally identify the geometric structure sought for. The following section describes this topic in detail.

6.6   Initial Singularity and Horizon

The big bang must coincide with the creation of a substratum of fundamental observers. By virtue of the cosmological principle, the distance vector $r(t)$ of a galaxy from one of the fundamental observers is a function of cosmic time τ, according to the relation:

$$r(t) = R(\tau)\boldsymbol{\xi} \tag{6.31}$$

where $R(\tau)$ is the scale distance (which we shall assume to be dimensionless) and $\boldsymbol{\xi}$ is the position vector of the galaxy in terms of comoving coordinates, independent of τ. We shall assume that $R(0) = 0$.

As we know, the de Sitter group is isomorphous to the group of rotations of the sphere in Euclidean 5-dimensional space. The equation of this hypersphere in projective coordinates is the following:

$$(\underline{x_0})^2 + (\underline{x_1})^2 + (\underline{x_2})^2 + (\underline{x_3})^2 + (\underline{x_5})^2 = r^2. \tag{6.32}$$

With a Wick rotation of the time coordinate $\underline{x_0}$, this equation becomes:

$$(\underline{x_0})^2 - (\underline{x_1})^2 - (\underline{x_2})^2 - (\underline{x_3})^2 + (\underline{x_5})^2 = r^2, \tag{6.33}$$

which is precisely the special case of (6.10) which corresponds to PSR.

As a first step, the canonical extension of (6.33) comprising (6.31) must be determined. This extension is obviously:

$$(\underline{x_0})^2 - R^2(\tau)[(\underline{x_1})^2 + (\underline{x_2})^2 + (\underline{x_3})^2] + (\underline{x_5})^2 = r^2. \tag{6.34}$$

As can easily be seen, by substituting into (6.34) the ordinary scale distance $R(\tau) \equiv 1$ we are back to (6.33) and therefore to PSR. Equation (6.34) takes on a physical meaning if τ is defined as a continuous monotonic function of $\underline{x_0}$. If the offset is chosen such that $x_0 = 0$ when $\tau = 0$, the big bang comes to be represented by the intersection of Eq. (6.4) with the hyperplane $x_0 = 0$, that is, since $R(0) = 0$, by the two equation points $(x_5)^2 = r^2$. These two points are the points of intersection of the axis x_5 with the 5-sphere (6.32).

At the time of the big bang, space as it is commonly understood is constituted by the three-dimensional hypersurface contained in the hyperplane $x_0 = 0$, perpendicular to the x_5 axis, tangent to the hypersphere (6.32) in the point $x_5 = r$. Thus the choice of the x_5 axis, which can be made in ∞^3 ways, determines the choice of a three-dimensional space, which is the *private space* of the fundamental observer adopting the x_1, x_2, x_3 coordinates. The metric of this space is described in the second, third and fourth term of (6.34); we are dealing with a space initially constituted by one single point (the fundamental observer himself), which then expands whilst maintaining its centre in the observer, in accordance with (6.31).

A rotation of the x_5 axis in the $x_0 = 0$ hyperplane is equivalent to a constant-time spatial translation, carried out at the instant of the big bang, which leads from one fundamental observer to another. This result differs radically from the traditional view of the big bang. In this view, the big bang is constituted by a single point-event into which all the fundamental observers converge. The big bang described by (6.34), on the other hand, is a three-dimensional manifold, in which ∞^3 fundamental observers can be selected, each with his own private space in which (6.31) is valid; the ξ coordinates are the comoving coordinates of the galaxy in this space.

At the time $\tau = 0$ there are, therefore, infinite distinct fundamental observers, *linked to each other by a global coordinate transformation group, which is a subgroup of the de Sitter group*. Equation (6.31) concerns *one* of these observers, and so does (6.34). Weyl's postulate on geodesics is therefore satisfied in each individual private space. In this vision, the big bang does not create space but takes place in a pre-existing (5-dimensional) space. As can be seen in (6.34), which is a special case of (6.27) corresponding to the cosmological problem, the big bang is simultaneous with the appearance of an absolute represented precisely by (6.34).

We had required that x_0 and τ be linked by a continuous monotone function; this condition is satisfied if it is assumed that on the manifold (6.34) the set of observation point-events corresponding to a same cosmic time is given by the intersection with the x_0 = constant hyperplane. The distance of these points from the x_0 axis is clearly $[r^2 - (x_0)^2]^{1/2}$.

Let us therefore consider the contemporaneousness plane of one of the point-events in question, say P. This hyperplane will form an angle φ with the big bang hyperplane whose equation is $x_0 = 0$. The tangent of this angle will be $x_0/[r^2 - (x_0)^2]^{1/2}$.

On the four-dimensional hyperplane tangent to the manifold (6.4) in P (P's private spacetime) the big bang will thus be projected geodesically at a chronological distance τ from P such that $\tau/t_0 = \mathrm{tg}(\varphi)$. The relation sought between x_0 and τ is therefore:

$$\tau = t_0 x_0 / [r^2 - (x_0)^2]^{1/2}. \tag{6.35}$$

It can be noted that τ increases indefinitely for $x_0 \to r$. This limit corresponds to the situation $\varphi \to \pi/2$, in which the geodesic projection of the big bang flattens out onto the de Sitter horizon of P's past light cone.

The cosmic time (6.35) is *kinematic*, in the sense that it includes the effects of the purely kinematic de Sitter expansion. As we have seen, the elimination of these effects requires Milne's regraduation, with the passage to the new *dynamic* (or *atomic*) cosmic time:

$$\tau' = t_0 + t_0 \ln\left(\frac{\tau}{t_0}\right). \tag{6.36}$$

The correct form of this regraduation which eliminates the paradoxes highlighted in the previous section is the following [12]:

$$\begin{aligned} \tau' \to \tau' &= \left[t_0 + t_0 \ln\left(\frac{\tau}{t_0}\right)\right] - \left[t_0 + t_0 \ln\left(\frac{t_{BB}(\tau)}{t_0}\right)\right] \\ &= t_0 \ln\left(\frac{\tau}{t_0}\right) - t_0 \ln\left(\frac{t_{BB}(\tau)}{t_0}\right). \end{aligned} \tag{6.37}$$

Specifically, it occurs that:

(1) for $\tau = 0$, $\tau' = -\infty$; the De Sitter horizon does not belong to GR spacetime;
(2) for $\tau = t_{BB} = t_0$, i.e., at the big bang, $\tau' = 0$ as required;
(3) only the $\tau \geq t_{BB}(\tau)$ values correspond to real physical phenomena;
(4) the property $\tau' = t_0$ for $\tau = t_0$, typical of eq. (6.36), is recovered in the special case $t_{BB}(t_0) = t_0/e$.

Finally, the metric coefficients of the individual private space can be deduced from (6.34), bearing in mind that it is a special case of (6.27):

$$\gamma_{0A} = \delta_{0A}; \quad \gamma_{5A} = \delta_{5A}; \quad \gamma_{ij} = -R^2(\tau)\delta_{ij}. \tag{6.38}$$

At this point, by adopting an appropriate expression for the projective energy tensor and assigning an equation of state, (6.25) can be solved and the scale distance $R(\tau)$ thus be determined. Here we summarize the results relating to a perfect fluid with density and pressure depending only on cosmic time; the interested reader is reminded to the original article [12] for details.

Firstly, the scale distance $R(\tau)$ is that which corresponds to the Fridman cosmological model, having a spatial curvature index $k = 0$, *without a cosmological term*. The solution to (6.25) is the metric with a Euclidean spatial section (in physical coordinates after applying the limit $r \to \infty$):

$$ds^2 = c^2 d\tau^2 - R^2(\tau)dx_i \, dx^i. \tag{6.39}$$

If regraduation (6.37) is now applied, the scale distance becomes that which corresponds to a Fridman model (in the new *atomic* cosmic time) with $k = 0$ and cosmological term $\lambda = 4/(3t_0^2)$. The passage to atomic time, consistent with General Relativity, thus causes *the cosmological term to appear kinematically.*

The relation between atomic time and kinematic time is expressed by one of the two equivalent forms [20]:

$$\frac{d\tau'}{d\tau} = \frac{1}{\sqrt{1 + \frac{3}{4}\lambda\tau^2}} = \frac{1}{\sqrt{1 + \left(\frac{\tau}{t_0}\right)^2}}, \tag{6.40}$$

$$\tau' = t_0 \, \text{arcsin h}\left(\frac{\tau}{t_0}\right). \tag{6.41}$$

This same relation is also presented, in a PSR context, in [21], Eq. (6.35).

6.7   Archaic Universe

The change of geometry required in Sect. 6.5 is therefore that which, at the same time as the big bang happens, causes the passage from the single hypersphere (6.32) to a multiplicity of spaces (6.34). At the instant $\tau = 0$, every *private* space (6.34) is reduced to a single point-event, which is one of the points belonging to the intersection of the $x_0 = 0$ hyperplane with the hypersphere (6.32). For the sake of brevity, this intersection from now on will be called the hypersphere's *equator*. The equator therefore constitutes the *public* space of the initial point-events coming out from the big bang.

As required in Sect. 6.5, the geometry (6.32) is atemporal, in the sense that the signature of the x_0 axis is the same as for the x_1, x_2, x_3 axes and therefore a true time axis does not exist. Time as it is commonly understood does not yet exist at this stage, which is that of a 4-space constituted by the surface of the 5-sphere (6.32). We can say, more correctly, that x_0 is a precursor of cosmic time, to which it comes to be connected through (6.35) after the big bang. The change of geometry is thus firstly associated with the emergence of cosmic time. We note that in Einstein's $r \rightarrow \infty$ limit, (6.35) becomes $\tau = x_0/c$ and the reference to the hyperspherical pre-space (6.32) is thus concealed.

With the passage to (6.34) the metric signature changes. The new coordinates allow of a light cone having its vertex in the observer, who can thus begin to "see" the surrounding world, recognizing it as his *environment*. In other words, the birth of the (fundamental) observer coincides with that of the possibility of his being distinguished from an external environment of signals received (the observer's past) or sent (the observer's future). Besides the signature change, a scale reduction by a factor $R(\tau)$ is also applied.

It is evident that these two transformations (of signature and of scale) *cannot correspond to a dynamic process*, i.e., to an ordinary physical transformation in time and in space. This would be impossible, on the other hand, since at the instant of this transformation time does not yet exist nor does space, as it is commonly understood. No dynamics exist, e.g. as described by a Lagrangian, which produce such a process. The relation between the public pre-space (6.32) and private spaces (6.34) is, rather, a relation of logical, archetypal origin. For this reason, the (6.32) pre-space can be defined as an "archaic Universe" [22]. The archaic Universe can be seen as the framework of conditions within which the private Universes of the various fundamental observers develop; indeed, within this very framework they can be recognized as different representations of one and the same Universe. Different points of view on a single reality.

The archaic Universe hypothesis makes possible a different approach to quantum cosmology. To understand this topic, consider that the relation (6.35), which makes the projective coordinate $\underline{x}_0$ a precursor of time, assigns this coordinate with a special role. Licata and Chiatti [12] have hypothesized that the archaic Universe was populated only by virtual quantum processes, quantum fluctuations originating on the equator ($\underline{x}_0 = 0$) and ending on the various $\underline{x}_0 =$ constant hyperplanes. The cosmological principle is thus no more than the assertion of the perfect homogeneity (apart from statistical fluctuations) of this archaic background of pre-big bang fluctuations. The fluctuation which manages to end at a given value of $\underline{x}_0$ is similar to a thermal fluctuation of energy kT which manages to reach the peak of the energy barrier $\hbar c / \underline{x}_0$. In other words, it is as if $\underline{x}_0$ were a measure of inverse absolute temperature:

$$x_0 = \hbar c / kT \qquad (6.42)$$

The $\underline{x}_0$ axis is therefore an axis of temperatures, and the physics of the archaic Universe is in certain respects similar to a form of thermal physics; the change of signature (Wick rotation) converts the inverse temperature into cosmic time. The background of fluctuations is a sort of pre-vacuum, distinct from the vacuum as ordinarily understood as it possesses energy. The hypothesis is that this form of pre-vacuum is unstable for fluctuations with $T \leq T_0$, i.e., which end at a "chronological" distance from the equator that is greater than $\theta_0 = \hbar / kT_0$. This means that the fluctuations belonging to this "tail" are instantly converted into *real quantum processes*. Their energy is converted into matter—which begins to interact with itself—and this conversion "empties" the pre-vacuum which is converted into ordinary vacuum. The ignition of the interactions also implies the ignition of gravitation. Starting from this moment, therefore, (6.25) must be applied and this leads to (6.44). The nucleation of matter and the inception of private spaces constitute the phenomenon called the big bang.

Licata and Chiatti explain the instability within the framework of the transactional approach to the understanding of quantum phenomena [12, 23] and hypothesize that T_0 is the Hagedorn temperature; if this is so, the entire archaic Universe phase corresponds to an extension—measured in x_0/c—of 10^{-23} s. The fact that the entire process is non-dynamic implies that the archaic Universe is an archetypal condition that can be reached even today, in processes in which the temperature T_0 is reinstated. Thus, the high-energy hadronic collisions with temporary formation of quark-gluon plasma (QGP) constitute so many current-day "returns" to that condition.

It can be noted that the scenario presented constitutes a quantum cosmology in a highly different sense from the traditional one. The archaic Universe is not small: its radius r measures approximately 1.5×10^{10} light years. Overall, the conjecture can be considered a sort of non-dynamic reformulation of the Hartle-Hawking argument [24, 25].

6.8 The Origin of Inertia

It can be shown that by describing the fluctuations of the archaic Universe with evanescent waves [26] of the type:

$$\Psi = \Psi_0 \exp\left[\pm j \frac{\sqrt{2mE}}{\hbar} x - \frac{E}{kT}\right] \tag{6.43}$$

where $j = \sqrt{-1}$, the transformation [induced by Eq. (6.32)$\rightarrow$(6.34)]:

$$t \rightarrow \frac{j\hbar}{kT} \tag{6.44}$$

transforms such waves into plane waves:

$$\Psi = \Psi_0 \exp\left[\pm j \frac{\sqrt{2mE}}{\hbar} x - j \frac{Et}{\hbar}\right] \tag{6.45}$$

which satisfy the customary dispersion relation (the classical one, in the example) and are associated with particles having kinetic energy E, mass m and impulse $(2mE)^{1/2}$. This is possible because (6.43) does not contain imaginary phase factors in T. In archaic space it represents a wave at rest in the frame of reference in which the wave equation that describes it applies; such a wave vanishes as T decreases.

In other words, the physical processes that populated the archaic Universe were fluctuations and therefore a privileged frame of reference had to exist in which the velocity of the motions involved had to be null. In such a frame, motion was exclusively temporal and the archaic equivalent of Heisenberg's indetermination principle was valid. Equation (6.32) is assumed to be written in this reference, and the various fundamental observers thus originate from the transformation (6.32)$\rightarrow$ (6.34) formulated in this frame. Naturally, after this transformation, archaic space ceased its function, so to speak, and the background that defined a privileged frame of reference disappeared. But the memory of the privileged frame was inherited as the selection of a particular substratum of fundamental observers, with respect to which the average peculiar motion of matter is null.

The non-peculiar motion of matter with respect to each given fundamental observer is established, on the other hand, by the cosmological principle, and it is the cosmic recession (6.31). The observer sees the galaxies move farther away from him according to a velocity-distance law which is the same in all directions and is the same for all observers at a given cosmic time. For a universe void of matter, in which PSR therefore applies and in which there is no expansion of space, such a law would be the well-known form of Hubble's law relating only to the kinematic effect of the de Sitter space curvature [6, 13]:

$$V = H(t)x, \qquad (6.46)$$

it being that $H(t) = H/(1 + t/t_0)$, $-t_0 \leq t \leq 0$, $H = 1/t_0$. In Einstein's limit $t_0 \rightarrow \infty$ one would therefore have $V \equiv 0$; i.e., the "system of fixed stars" would be obtained again. The fundamental observer's frame of reference corresponds, therefore, to that which in the limit of classical physics or SR is the fixed-stars frame of reference.

6.9   Dark Matter

Let us now return to the projective metric solution to the cosmological problem, Eq. (6.38). and we note that, apart from the presence of the scale distance $R(\tau)$, it coincides with the customary PSR metric. The passage to the physical metric must be carried out in accordance with substitutions (6.14)–(6.16). The result is that physical spacetime admits of a hyperbolic spatial section, basically coinciding with that of PSR if the scale distance is omitted. This result naturally does not contradict the Euclidean spatial section of (6.39), which only represents its Einstein limit $r \rightarrow \infty$. The salient point is that *the spatial section is Euclidean only in the Einstein limit*. This conclusion represents a substantial difference from Fridman cosmology which, instead, associates a *globally* Euclidean spatial section with the $k = 0$, $\lambda > 0$ model. We saw before that the passage to cosmic time of GR entailed the appearance of the cosmological term and of dark energy. What happens if not only GR time is adopted but also its system of Euclidean spatial coordinates? To understand this topic, let us first consider the PSR context and then introduce the required changes related to the variation of scale distance in cosmic time [20].

In PSR, an inertial observer O sees a second inertial observer P, located at spatial distance x and at chronological distance t, receding at velocity (6.46). The apparent acceleration of P can be obtained by deriving (6.46) with respect to $-t$. In O's contemporaneousness space (i.e., for $t = 0$) this acceleration is:

$$a_{recession} = \frac{c}{t_0}\frac{x}{r} = \frac{x}{t_0^2};$$ (6.47)

it is maximum for $x = r = ct_0$ and in this case it is equal to c/t_0. Thus, in accordance with O's point of view, P's rest frame of reference is accelerated. This acceleration is the expression of the hyperbolic geometry of O's contemporaneousness space.

Coordinating the events of O's contemporaneousness plane with a Euclidean system of coordinates means passing to a new representation in which this acceleration is null. In Newtonian cosmology terms [18], this is equivalent to assuming a density ρ of cosmic fluid such that the gravitational attraction exercised on P by the fluid sphere of radius $x = OP$ centred on O is exactly that required to offset the acceleration of P's recession motion. That is, it must be:

$$\frac{4\pi G\rho}{3}x = \frac{x}{t_0^2} \Rightarrow \rho = \frac{3}{4\pi Gt_0^2}.$$ (6.48)

where G is Newton's gravitational constant. The expansion of space can be taken into account by substituting ρ with ρR^3 and x with (x/R) in (6.48), in order to obtain a relation that is not dependent on cosmic time. One thus has:

$$\frac{4\pi G}{3}\left(\rho R^3\right)\left(\frac{x}{R}\right) = \frac{1}{t_0^2}\left(\frac{x}{R}\right) \Rightarrow \rho = \frac{3}{4\pi Gt_0^2 R^3};$$ (6.49)

but this is precisely the cosmic fluid density obtained by the solution of (6.25) [20]. It coincides with the well-known critical density.

One can thus maintain the following line of reasoning. The form of Eq. (6.34), in which curvature factors are absent, is imposed by the condition that for $R \equiv 1$ it must return (6.33) that is valid in PSR; this is a structural constraint. Equation (6.34) implies (6.38) which, when substitued into (6.25), give (6.39). Equation (6.39) does not represent the correct line element, but only its Einstein limit; it is compatible with a cosmic fluid density that corresponds to the critical one. However, the correct line element is that expressed by the hyperbolic metric induced on the physical coordinates, and in these coordinates the fundamental observers are mutually accelerated. The gravitational effect of cosmic fluid does *not* offset this acceleration, and the real density of this fluid must therefore be less than the critical one. The difference between the critical density and the real density appears as the cosmological background of "dark matter". In other words, it is plausible that this background is a projective effect, closely associated with the cosmological term.

Assuming that this explanation of the cosmological background of dark matter is correct, it cannot in any case apply to dark matter thickened into structures (galaxies and clusters). The perturbation of the motion of these structures is indeed a local phenomenon which can in no way be connected with projective effects or the particular choice of global frames of reference. It represents a real physical effect which requires an explanation based on new physics. In the perspective set forth here, the problem of the cosmological background of dark matter must therefore be distinguished from that of "local" dark matter.

The problem of "thickened" dark matter and of its relation with PGR is an entirely open one. One possibility [20, 27] is that the dark matter thickenings are fossils of the archaic era. To illustrate this concept, let us consider once again the privileged archaic frame of reference described in Sect. 6.8. It is not strictly necessary that the motions within this system be reduced exactly to translations along the inverse temperature axis such as to constitute the equivalent of rest states. More generally, zero-mean oscillatory motions, possibly multiperiodic ones, are also admissible. One can therefore hypothesize the existence, in this phase, of spatial regions in which the equivalent of free motion is constituted by oscillations (in the inverse temperature) around a point of equilibrium. The $(6.32) \rightarrow (6.34)$ transformation thus converts these regions into regions where the inertia principle is violated, in that free motion becomes a periodic motion that is accelerated around a centre. These regions thus constitute a sort of granular structure within the cosmic fluid, or actual "molecules" of fluid. While the non-granular component of fluid constitutes an inertial structure within which (6.25) is exactly valid, inside a given molecule the presence of a "natural" field of accelerations corresponds, from the standpoint of the application of (6.25), to the presence of a density of "anomalous" matter. This would be the thickened dark matter. Naturally, after the big bang the various "molecules" are superposed and their acceleration fields compose themselves, giving rise to a nearly null mean. But, as the expansion of space separates the molecules, their structure becomes increasingly defined until, approximately 7 billion years after the big bang [27], they are completely separated. In this first phase of the history of the Universe the molecules thus act as attractors of condensing matter, and therefore as privileged centres for structure formation. The vorticity of the structures (galaxies and clusters) may have been originated by the fall of matter into attraction basins in the process of separation, though this idea has not yet been substantiated by detailed calculations.

A quantitative development of these ideas [27, 28] leads to the astonishing conclusion that the de Sitter time t_0 governs the rotation of galaxies. The quantization of "archaic" periodic motions implies that the accelerations associated with them, and which reappear inside the molecules after the big bang, are of the order of c/t_0. This acceleration, independent of cosmic time, therefore measures the extent of the violation of the inertia principle inside the molecules, in the sense that a free material point inside a molecule undergoes an acceleration of this extent towards the centre of the molecule. The condition of equilibrium on the rotating motion of a galaxy located within the molecule will thus contain t_0. This cosmological parameter will thus co-determine the galactic rotation.[3] The most important consequence is given by the existence of the following scaling law:

---

[3]In fact, the maximum accelerations recorded in the rotation of galaxies are of the order of c/t_0; it is possible to formulate semi-empirical quantitative models of spiral rotation based on these ideas, which agree very well with observations [27, 28].

$$\mu\xi = \frac{3c}{4\pi G t_0}.$$ (6.50)

where μ is the density of dark matter present in a structure and ξ is the radius of its distribution. This law is verified on all size scales, from dwarf galaxies to giant spirals [28]. It is worthy of note that the second member of (6.50) represents the product of the critical density (evaluated at the moment when the radius of the Universe is equal to the de Sitter radius) by r.

6.10   The Dominance of Matter

The reason for the dominance of ordinary matter in our Universe remains an enigma also within the PGR approach [26]. The equator of the Arcidiacono hypersphere corresponding to an infinite absolute temperature is the origin of all the "archaic" quantum fluctuations. Those which end at a temperature equal to the creation energy of nucleons are converted into real particles which emerge as the "big bang". Since from this moment on the temperature is too low for there to be interconversion of baryons and antibaryons, the dominance of matter must be established beforehand, during the "archaic era". However, in the archaic era only the quantum pre-vacuum is present, which as such is expected to be symmetrical to the utmost. The enigma, therefore, reappears in this approach, as well.

It must be borne in mind, however, that in the approach presented here it is possible to exploit a symmetry which is lacking in more conventional cosmology. Our Universe develops from one of the two semispheres into which the infinite-temperature equator divides the hypersphere (6.32). The semisphere is converted into PGR spacetime by means of transformation (6.32)→(6.34).

The other semisphere does not play any role and can be considered as a simple mathematical artefact [it corresponds to the choice of a negative sign instead of a positive sign in the definition of the projective coefficient A of the metric (6.11c)]. However, it is also possible that this semisphere corresponds to a second archaic universe that is specular with our own. If the archaic fluctuations exiting from the equator towards this second semisphere corresponded solely to antimatter particles and those exiting towards our semisphere corresponded solely to particles of ordinary matter, the symmetry of the original pre-vacuum would be complied with.

Thus, in the second semisphere, by means of the same mechanisms seen above, a Universe would be developed in which antimatter would be dominant. And this would take place whilst maintaining the symmetry of the archaic quantum pre-vacuum. These two "mutually specular Universes" would be separated by the equator, and would therefore be causally unconnected, albeit contiguous. Their common origin would be the archaic quantum pre-vacuum.

The mechanism based on which ordinary matter and antimatter would seem to separate along the equator remains enigmatic. A particle differs from its antiparticle by the sign of the charges with which they couple with the various fields. On the other hand, if one admits that the various interaction fields present in the archaic era satisfy the Gauss theorem, one easily sees that the charges must be null because of the fact that space is closed. One can therefore presume that the charges make their appearance only at the time of the big bang, when the particles become real and therefore capable of real interaction and space becomes open. Only at that time does each particle acquire its own charge.

If one assumes that the sign of the charges is defined by the semisphere in which the particle becomes real (i.e., by the fact that it appears at the time of the big bang in our Universe or, alternatively, at the time of the anti-big bang in the anti-Universe) one does indeed obtain the required separation ab initio of matter and antimatter. The sign of the charge would in other words be defined by the direction of the timeline emerging from the equator along which the particle materialized, a result which could in some way be connected with the CPT theorem. Though the entire conjecture does not appear open to verification by observation, it nevertheless allows to explain the dominance of ordinary matter without introducing special initial conditions, at the same time preserving the greatest symmetry of the initial state.

In line of principle, another possibility is that separation ab initio has never happened, but instead a quantity A (B) of matter (antimatter) has been injected in our Universe at the big bang, and a quantity A (B) of antimatter (matter) has been injected in the anti-Universe at the anti-big bang. In this case, the condition A > B assures the dominance of matter in our Universe, which is however balanced by the dominance of antimatter in the anti-Universe. The total annihilation of antimatter in the early days of our Universe should have released a large amount of radiation pressure [29].

References

1. Arcidiacono, G.: De Sitter universe and projective relativity. Coll. Math. **19**, 51–72 (1968)
2. Arcidiacono, G.: De Sitter universe and astrophysics. Coll. Math. **25**, 295–317 (1974)
3. Arcidiacono, G.: Cartan spaces and unitary theories. Coll. Math. **16**, 149–168 (1964)
4. Arcidiacono, G.: The curvature-torsion cartan tensor and the projective general relativity. Coll. Math. **34**, 95–107 (1984)
5. Veblen, O.: Projective relativitätstheorie. Springer, Berlin (1933)
6. Arcidiacono G.: The theory of the universes (vol. II). Di Renzo, Roma (2000)
7. Arcidiacono, G.: Projective general relativity and the vectorial-tensorial field. Coll. Math. **34**, 197–206 (1984)
8. Arcidiacono, G.: Projective general relativity and the scalar-tensorial field. Coll. Math. **34**, 115–129 (1984)
9. Arcidiacono, G.: The projective scalar-tensorial gravitational field and the conformal transformation of the metric. Hadr. Journ. **11**, 287 (1988)
10. Singh, T., Singh, G.P.: Non-static cylindrically symmetric fields in general projective relativity. Astrophys. Space Sci. **163**, 41–48 (1990)

11. Singh, T., Singh, G.P.: Liouville space-time in general projective relativity. Astrophys. Space Sci. **173**, 227–232 (1990)
12. Licata, I., Chiatti, L.: Archaic universe and cosmological model: "big-bang" as nucleation by vacuum. Int. Jour. Th. Phys. **49**(10), 2379–2402 (2010)
13. Chiatti, L.: The fundamental equations of point, fluid and wave dynamics in the de Sitter-fantappie-arcidiacono projective relativity theory. El. Journ. Theor. Phys. **23**(7), 259–280 (2010)
14. Kerner, H.E.: Proc. Natl. Acad. Sci. U.S.A. **73**, 1418–1421 (1976)
15. Arcidiacono, G.: De Sitter universe and mechanics. Coll. Math. **20**, 231–256 (1969)
16. Arcidiacono, G.: The klein model in cosmology. Coll. Math. **25**, 159–184 (1974)
17. Chiatti, L.: Fantappié-Arcidiacono theory of relativity versus recent cosmological evidences: a preliminary comparison. El. Journ. Theor. Phys. **4**, 17–36 (2007)
18. Bondi, H.: Cosmology. Cambridge University Press, Cambridge (1961)
19. Milne, E.A.: Kinematic Relativity. Clarendon Press, Oxford (1948)
20. Chiatti, L.: De Sitter relativity and cosmological principle. Open Access Astr. Journ **4**, 27–37 (2011)
21. Guo H.Y., Huang C.G., Xu Z., Zhou B.: On special relativity with cosmological constant; arXiv:hep-th/0403171v1
22. Licata, I., Chiatti, L.: The archaic Universe: big bang, cosmological term and the quantum origin of time in Projective Cosmology. Int. Journ. Theor. Phys. **48**(4), 1003–1018 (2009)
23. Chiatti L.: The Transaction as a quantum concept. In: Licata I. (ed) Space-time geometry and quantum events, pp. 11–44. Nova Science, New York (2014)
24. Licata I.: Transaction and non locality in quantum field theory. In: EPJ web of conferences 70 00039 (2014). doi:10.1051/epjconf/20147000039
25. Licata, I.: Universe without singularities, a group approach to de Sitter cosmology. El. Journ. Theor. Phys. **10**, 211–224 (2006)
26. Chiatti, L.: A possible mechanism for the origin of inertia in De Sitter-Fantappié-Arcidiacono projective relativity. El. Journ. Theor. Phys. **9**, 11–26 (2012)
27. Chiatti, L.: Cosmos and particles, a different view of dark matter. Open Access Astr. Journ. **5**, 44–53 (2012)
28. Chiatti, L.: Dark matter in galaxies: a relic of a pre-big bang era? Quant. Matter J. **3**(3), 284–288 (2014)
29. Alfvén H.: Worlds-antiworlds: antimatter in cosmology. W. H. Freeman and Company, S. Francisco and London (1966)

Appendix A

Curvature and Torsion in PGR

Let us remember that, following the definition of Cartan, any Riemann manifold is associated with an infinite family of Euclidean spaces tangent to it in each of its P points. These infinity spaces are joined by a connection law and are individuated by a holonomy group. By introducing a local coordinates system y^i and a linear forms, ω^i, of differential dy^i we can write $ds^2 = \omega_s \omega^s$. If we consider, on the tangent space to a point P, four orthogonal vectors e_i we have

$$\begin{cases} dP = \omega^i e_i \\ de_i = \omega_i^k e_k \\ e_i e_k = \delta_{ik} \end{cases} \qquad (A.1)$$

where $\omega_k^i = \gamma_{ks}^i \omega^s$ and γ_{ks}^i are the Ricci rotation coefficients. If the point P and the associated reference frame describe a closed infinitesimal cycle on the tangent space, in general the vector e_i' doesn't coincide with e_i and the cycle is open. It can be closed through a translation Ω^i and a rotation Ω_k^i on the tangent space and we have

$$\begin{cases} \Omega^i = d\omega^i + \omega_s^i \wedge \omega^s \\ \Omega_k^i = d\omega_k^i + \omega_s^i \wedge \omega_k^s \end{cases} \qquad (A.2)$$

where Ω^i is the torsion and Ω_k^i is the curvature. To develop the projective general relativity we have to introduce a 5-dimensional Riemann manifold which allows as holonomy group the de Sitter one, isomorphic to the 5-dimensional rotations group, and the gravitation equations are

$$R_{AB} - \frac{1}{2} g_{AB} R = \chi T_{AB} \quad (A, B = 0, 1, 2, 3, 4) \qquad (A.3)$$

© The Author(s) 2017
I. Licata et al., *De Sitter Projective Relativity*, SpringerBriefs in Physics,
DOI 10.1007/978-3-319-52271-5

where g_{AB} are the coefficients of the five-dimensional metric. We immediately understand, from how much we have said above, that, in projective general relativity, is fundamental the geometry of projective connections and therefore let us remember the following concepts.

If we have a differentiable manifold M and its symmetric connection θ, the curvature tensor in local coordinate is

$$K^{\alpha}_{\beta\gamma\delta} = \frac{\partial \theta^{\alpha}_{\beta\delta}}{\partial x^{\gamma}} - \frac{\partial \theta^{\alpha}_{\beta\gamma}}{\partial x^{\delta}} + \sum_{\sigma} \left(\theta^{\alpha}_{\sigma\gamma}\theta^{\sigma}_{\beta\delta} - \theta^{\alpha}_{\sigma\delta}\theta^{\sigma}_{\beta\gamma} \right). \tag{A.4}$$

The Ricci tensor of connection is

$$K_{\beta\delta} = \sum_{\alpha} K^{\alpha}_{\beta\alpha\delta} \tag{A.5}$$

By setting

$$A_{\beta\delta} = \frac{1}{2}\left(K_{\beta\delta} - K_{\delta\beta} \right) = \frac{1}{2}\sum_{\alpha} \left(\frac{\partial \theta^{\alpha}_{\alpha\delta}}{\partial x^{\beta}} - \frac{\partial \theta^{\alpha}_{\alpha\beta}}{\partial x^{\delta}} \right) \tag{A.6}$$

the following tensor

$$\begin{aligned} W^{\alpha}_{\beta\gamma\delta} = K^{\alpha}_{\beta\gamma\delta} &- \frac{2}{n+1}\delta^{\alpha}_{\beta}A_{\gamma\delta} - \frac{1}{n-1}\left(\delta^{\alpha}_{\gamma}K_{\beta\delta} - \delta^{\alpha}_{\delta}K_{\beta\gamma} \right) \\ &+ \frac{2}{n^2-1}\left(\delta^{\alpha}_{\gamma}A_{\beta\delta} - \delta^{\alpha}_{\delta}A_{\beta\gamma} \right) \end{aligned} \tag{A.7}$$

is said projective curvature tensor. Two such connections are projectively equivalent if they define the same geodesics up to parametrization. A n-dimensional differentiable manifold M with symmetric connection θ is said locally projectively flat if $\forall x \in M$ there is a neighborhood U and a diffeomorphism from U to an open of R^n which transforms the images of geodesics of θ, contained in U, into straight lines. Let us remember that M is locally projectively flat if and only if the projective curvature tensor is identically zero and, if the Ricci tensor is symmetric, the manifold M is locally projectively flat if and only if the curvature tensor can be written with the following relation

$$K^{\alpha}_{\beta\gamma\delta} = \frac{1}{n-1}\left(\delta^{\alpha}_{\gamma}K_{\beta\delta} - \delta^{\alpha}_{\delta}K_{\beta\gamma} \right). \tag{A.8}$$

Given a Riemannian manifold M and u and v, two linearly independent tangent vectors at the same point x_0, we can define

$$K(u,v) = \left[\frac{\sum R_{\alpha\beta\gamma\delta} u^\alpha v^\beta u^\gamma v^\delta}{\sum \left(g_{\alpha\gamma} g_{\beta\delta} - g_{\alpha\delta} g_{\beta\gamma} \right) u^\alpha v^\beta u^\gamma v^\delta} \right] (x_0). \qquad (A.9)$$

It can be shown that $K(u,v)$ depends only on the plane spanned by u and v and it is called sectional curvature. In a Riemannian manifold the relation (A.8) can be written:

$$R^\alpha_{\beta\gamma\delta} = \frac{1}{n-1} \left(\delta^\alpha_\gamma R_{\beta\delta} - \delta^\alpha_\delta R_{\beta\gamma} \right) \qquad (A.10)$$

and setting as usual

$$R_{\alpha\beta\gamma\delta} = \sum_\rho g_{\alpha\rho} R^\rho_{\beta\gamma\delta} \qquad (A.11)$$

we can write

$$R_{\alpha\beta\gamma\delta} = \frac{1}{n-1} \left(g_{\alpha\gamma} R_{\beta\delta} - g_{\alpha\delta} R_{\beta\gamma} \right) \qquad (A.12)$$

The condition

$$R_{\alpha\beta\gamma\delta} + R_{\beta\alpha\gamma\delta} = 0 \qquad (A.13)$$

becomes

$$g_{\alpha\gamma} R_{\beta\delta} - g_{\alpha\delta} R_{\beta\gamma} + g_{\beta\gamma} R_{\alpha\delta} - g_{\beta\delta} R_{\alpha\gamma} = 0 \qquad (A.14)$$

and therefore

$$\sum_{\alpha,\gamma} g^{\alpha\gamma} \left(g_{\alpha\gamma} R_{\beta\delta} - g_{\alpha\delta} R_{\beta\gamma} + g_{\beta\gamma} R_{\alpha\delta} - g_{\beta\delta} R_{\alpha\gamma} \right) = 0 \qquad (A.15)$$

that is

$$n R_{\beta\delta} - R_{\beta\delta} + R_{\beta\delta} - R g_{\beta\delta} = 0 \qquad (A.16)$$

and we get

$$R_{\beta\delta} = \frac{R g_{\beta\delta}}{n}. \qquad (A.17)$$

where $R = \sum_{\alpha,\gamma} g^{\alpha\gamma} R_{\alpha\gamma}$ is the scalar curvature. By replacing (A.17) in (A.10) we obtain

$$R^\alpha_{\beta\gamma\delta} = \frac{R}{n(n-1)} \left(\delta^\alpha_\gamma g_{\beta\delta} - \delta^\alpha_\delta g_{\beta\gamma} \right). \tag{A.18}$$

We can conclude saying that a Riemannian manifold is locally projectively flat if and only if the sectional curvature is constant. Therefore while in classical General Relativity the curvature tensor equal to zero means Minkowski spacetime, in Projective General Relativity curvature tensor equal to zero means de Sitter spacetime.

Appendix B
Interlude

Some Remarks About the Current Status
of Cosmological Problem

A comparing between the Friedman-Robertson-Walker (FRW) metric and the De Sitter Universe.

It is well-know that the major part of the theories of gravitation are obtained by modifying the Einstein-Hilbert action, adding scalar fields or curvature invariants. If we apply Einstein's equations

$$R_{\mu\nu} - \frac{1}{2}g_{\mu\nu}R = \frac{8\pi G}{c^4}T_{\mu\nu} \tag{B.1}$$

to the whole Universe, we find the relativistic cosmology, in which the cosmological principle can be postulated and a model of constant spatial curvature obtained. $R_{\mu\nu}$ is the Ricci tensor, R is the Ricci scalar and $T_{\mu\nu}$ is the stress-energy tensor. By assuming that matter consists of an ideal fluid we can write

$$T_{\mu\nu} = (p + \varepsilon)u_\mu u_\nu - pg_{\mu\nu} \tag{B.2}$$

where p is the pressure, ε the energy density and u_μ the velocity of the fluid respectively. Assuming homogeneity and isotropy, the Friedman-Robertson-Walker (FRW) metric describes the right geometry of a spacetime where the spatial curvature is constant

$$ds^2 = c^2dt^2 - a^2(t)\left[\frac{dr^2}{1 - kr^2} + r^2\left(d\theta^2 + \sin^2\theta d\varphi^2\right)\right] \tag{B.3}$$

in which (r, θ, φ) are the comoving coordinates, $a(t)$ is the scale factor and $k = 0, \pm 1$ the curvature constant. In such a context we obtain the relations:

© The Author(s) 2017
I. Licata et al., *De Sitter Projective Relativity*, SpringerBriefs in Physics,
DOI 10.1007/978-3-319-52271-5

$$\frac{\ddot{a}}{a} = \frac{-4\pi G}{3c^2}(\varepsilon + 3p) \tag{B.4}$$

$$\left(\frac{\dot{a}}{a}\right)^2 + \frac{kc^2}{a^2} = \frac{8\pi G}{3c^2}\varepsilon \tag{B.5}$$

$$\dot{\varepsilon} + 3\left(\frac{\dot{a}}{a}\right)(\varepsilon + p) = 0 \tag{B.6}$$

where a dot denotes a derivative with respect to the cosmic time t. A further equation has to be imposed in order to assign the thermodynamical state of matter:

$$p = \gamma\varepsilon = \gamma c^2 \rho \tag{B.7}$$

where γ is a constant ($0 \leq \gamma \leq 1$ for standard perfect fluid matter).

We have to pay close attention to the extrapolation of these equations to very far regions of spacetime, where they could be little more than a good model. As it is known, minor changes to the equations, while exhibiting all the classical verifications, produce completely different cosmologically-interesting solutions. It is useful to observe that even though cosmology accepts General Relativity as a definitive theory of gravitation, there are still some uncertain aspects due to a baffling pluralism. Possible universes are numerous and differ from each other substantially. Concerning the theory of the *big bang*, the primary difficulty is the presence of the initial singularity which brings up the problem of behaviour of matter when it is reduced to no dimensions with infinite density and temperature. The relativistic cosmology is unable to provide an explanation as to why the density of the universe should be so close to the critical value. In fact we have

$$\Omega - 1 = \frac{k}{H^2 a^2} \tag{B.8}$$

where Ω is the density parameter, H is the Hubble constant, k is the curvature, and a is the scale factor. We see that Ω will be much larger (or smaller) than 1, the greater will be $\left|\frac{k}{H^2 a^2}\right|$. This ratio does not remain constant during the expansion, since, while the numerator is constant, the denominator decreases and thus the ratio increases in absolute value. During the radiation era, we have $H \propto t^{-1}$ and $a \propto t^{1/2}$ and therefore $H^2 \propto t^{-2}$ and $a^2 \propto t$. Then $H^2 a^2 \propto t^{-1}$ with

$$\left|\frac{k}{H^2 a^2}\right| \propto t \tag{B.9}$$

This means that, during the radiation era, any difference of Ω from 1 increases proportionally with time. During the matter era, instead, we have $H \propto t^{-1}$ and $a \propto t^{2/3}$ and therefore $H^2 \propto t^{-2}$ and $a^2 \propto t^{4/3}$. Then $H^2 a^2 \propto t^{-2/3}$ with

$$\left| \frac{k}{H^2 a^2} \right| \propto t^{2/3} \tag{B.10}$$

This means that during the matter era any difference of Ω from 1 increases with time less rapidly than is the case during the radiation era. Therefore the Universe diverges from the flat case if $k \neq 0$ and the fact that it appears to be almost flat today means that Ω must have been very close to one in the early universe. Since the cosmological models tend to accentuate dramatically, during the expansion, any difference of Ω to 1, it is reasonable to ask why, given the current value of Ω, the universe has started its expansion in a state so close to a Euclidean situation.

In 1905, Einstein built his theory of Special Relativity on the assumption, experimentally verified, that the speed of light is a constant of nature. Its value does not vary, whatever the state of motion of the observer with respect to the light source. Consequently no material body can reach nor exceed the speed of light. This has important consequences on the processes of causality, since no causal interaction between two objects can be transmitted instantly, and the minimum time interval between the cause and its effect is closely related to the space between them and the speed of light c. The cosmic microwave background is observed to be extremely homogeneous and isotropic on large scales, with temperature fluctuations of $10^{-5} K$. This suggests that all regions of the sky were in casual contact at some time in the past, but is contradicted as follows. The horizon size is the distance light has travelled since the beginning of the universe and is given by

$$d(t) = a(t) \int_{t_1}^{t_2} \frac{dt}{a(t)} \tag{B.11}$$

which remains finite as $a(t_1) \to 0$ if $a > 0$. When the microwave background was formed the region in casual contact would have been approximately 0.09 Mpc. With the subsequent expansion this corresponds to a patch of the present microwave background subtending an angle of only $2°$.

The most principal problem is the singularity problem, and, according to Hawking-Penrose theorems, the appearance of singularity in cosmological solutions of general relativity is inevitable. Many physicists and cosmologists are inclined to believe that classical General Relativity must be revised in the case of extremely high energy densities, pressures, and temperatures. The singularity must mean for cosmology that the classical Einsteinian theory is inapplicable in the beginning of cosmological expansion of the universe. Finally recent observations of type Ia supernovae indicate that the universe is in an accelerating expansion phase and its geometry is flat. These observations give rise to the search for a field which can be responsible for accelerated expansion, and Dark Energy is the most popular way to explain these recent observations. The simplest Dark Energy candidate is the cosmological constant Λ introduced by Einstein to obtain a static universe but later on he himself rejected it. This vacuum energy density is equivalent to a perfect fluid obeying the equation of state $p_A = -\rho_A$. However, the nature and cosmological origin of Dark Energy still remain enigmatic at present, and it is not clear yet whether Dark Energy can be described by a cosmological constant which is independent of time or by dynamical scalar fields such as quintessence. Finally let us remember that inflation has been very successful in solving the problems in the standard big bang cosmology such as the horizon and flatness problems. When the universe is dominated by the material whose equation of state satisfies $p < -\rho/3$ the accelerating universe is realized. A natural candidate of matter with negative pressure necessary to drive inflation is a scalar field, so-called "inflaton." A homogeneous classical scalar field $\Phi = \Phi(t)$ is characterized by the energy density $\rho = \rho(\Phi)$ and the pressure $p = p(\Phi)$

$$\rho(\Phi) = \frac{1}{2}\dot{\Phi}^2 + V(\Phi) \tag{B.12}$$

$$p(\Phi) = \frac{1}{2}\dot{\Phi}^2 - V(\Phi) \tag{B.13}$$

where $V(\Phi)$ is the potential of the scalar field. From the Klein-Gordon equation, the equation of motion for the homogeneous scalar field is given by

$$\ddot{\Phi} + 3H\dot{\Phi} + V_\Phi = 0 \tag{B.14}$$

where $V_\Phi = \frac{dV}{d\Phi}$. In the scalar field dominant universe, the Friedmann equations become

$$H^2 = \left(\frac{\dot{a}}{a}\right)^2 = \frac{8\pi G}{3c^2}\left(\frac{1}{2}\dot{\Phi}^2 + V(\Phi)\right) \tag{B.15}$$

$$\dot{H} = \frac{-4\pi G}{c^2}\dot{\Phi}^2 \tag{B.16}$$

Inflation is realized when $\frac{1}{2}\dot{\Phi}^2 \ll V$.

Index

© The Author(s) 2017   105
I. Licata et al., *De Sitter Projective Relativity*, SpringerBriefs in Physics,
DOI 10.1007/978-3-319-52271-5

Printed in the United States
By Bookmasters